INTERNATIONAL PC

CONCISE POULTRY COLOUR GUIDE

NOTE:

The descriptions of breeds are in the order shown in the Breeds Index in the Preliminary pages.
The **General Index** is at the back of the book.

COLOUR PLATES ARE BETWEEN PAGES 48 & 49.

OTHER TITLES AVAILABLE OR IN PRODUCTION

A wide range of titles is available and a list is available on request. (Large SAE please). Examples are as follows:

Bantams : A Concise Guide J Batty

Bantams & Small Poultry, J Batty

Old & Rare Breeds of Poultry

J Butler

Understanding Modern Game

K J G Hawkey

Understanding Modern Game

J Batty & J P Bleazard

Poultry Houses & Appliances – A DIY Guide

Breeds of Poultry & Their Characteristics J Batty

Poultry Diseases under Modern Management G S Coutts

The Silkie Fowl J Batty

Domesticated Ducks & Geese, J Batty

Bantams as Layers, J Barnes

The Art of Cockfighting Arch Ruport

Concise Poultry Colour Guide, J Batty

Sebright Bantams, J Batty

Sussex & Dorking Fowls, J Batty

Old English Game Bantams, J Batty

Pekin Bantams, Margaret Gregson

CONCISE POULTRY COLOUR GUIDE

JOSEPH BATTY

Beech Publishing House
Station Yard
Elsted, Midhurst
West Sussex GU28 0JT

First published as *The Poultry Colour Guide* for the Centenary of the Poultry Club in 1977.

This 5th Edition 1996 in A5 format, but greatly enlarged on text. Ducks & Geese now covered in ***Ducks & Geese Colour Guide.***

ISBN 1-85736-210-1

Beech Publishing House
Station Yard
Elsted, Midhurst
West Sussex GU28 0JT

CONTENTS

La Fleche

An extraordinary bird with horns, which is a glossy green-black.

PREFACE

My thanks are offered to Charles Francis who painted the original paintings for *The Poultry Colour Guide.*

A decision was taken to change the format and enlarge the content because, in the light of the new knowledge available and the greater interest in genetics, it was felt that readers should be given more detail on how the colours evolved.

New breeds have also been included and the USA standards, as well as the British, were also consulted and appropriate notes are given.

BREEDS INDEX

1

THE POULTRY EVOLUTION

BRIEF HISTORY

Poultry have existed for a considerable period of time, exactly how long is not known with certainty. Estimates have ranged considerably, but it seems the race *Gallus* has existed for about 8 million years, and domestication of poultry occurred in China about 6,000 BC and in India around 2,000 BC.*

The present domesticated breeds go back hundreds of years for some, but are of relatively recent development for others, especially those created in Victorian times in the period of the so called Hen-fever era.

Bantams are probably the oldest known breeds, coming from many lands. In fact, the name itself is said to originate from a district in Java, from whence came the first recognized bantams in Britain.

The original bantam breeds in this country included the Pekin, and Sebright bantams which were developed about

* ***Genetics & Evolution of the Domestic Fowl,*** **Lewis Stevens, Cambridge, 1991. This book provides an excellent background for those who wish to consider in depth the evolution of the fowl.**

200 years ago*. However, they were certainly preceded by ***large*** fowl, Old English Game and Dorkings being the first domesticated types recorded. The Romans found the Dorkings in existence or brought them with them at the beginning of the Christianity period. There were also in existence the Old Kent Fowl (now merged with Sussex or Dorkings or both) and what became the Sussex breeds, both related to the Dorking. In Scotland there was the Scots Dumpie, a peculiar fowl of low stature, rather like a dwarf, and this appears to be of ancient origins; these belong to a group known as "Creepers" which, when bred together, have lethal genes.

Other breeds came into the country from far and wide. There were also some early breeds, known in the north of England, probably from Europe, and included a form of Hamburgh. These were found in the Yorkshire and Lancashire areas under the names of Moonies, Pheasant Fowl, and Bolton Greys**. The Poland was a similar long-standing breed which was on the continent for generations, having a distinctive crest and cavernous nostrils.

From the Mediterranean countries came the ***laying types of breed***. These are highly developed to the extent that they no longer carry out their normal and natural function of sitting on eggs and rearing their young. In this group appeared Anconas, Leghorns, Andalusians, Minorcas and Spanish. All very much geared to laying large numbers of eggs, some achieving records in the region of 300 per annum.

***See *Bantams & Small Poultry*, Joseph Batty, Elsted, 1996, which describes all the known breeds of bantams.**

**** Interestingly these appear to have been the nucleus for the first shows for standard breeds. *(Poultry Shows & Exhibiting,* J Batty)**

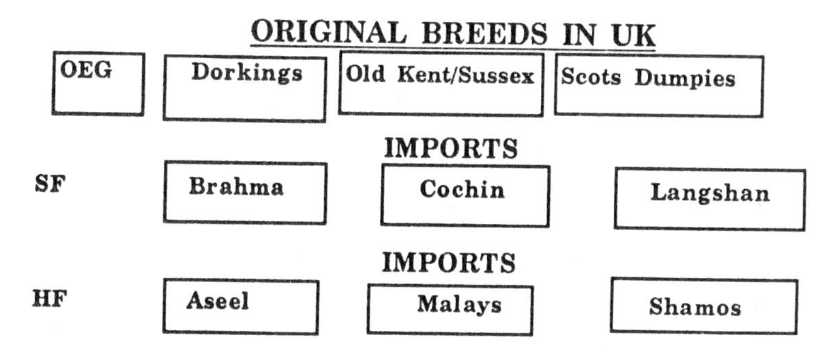

These were used as crosses to produce a variety of breeds, mainly heavy types; eg, Orpingtons, Wyandotte, Indian Game, Modern Game, etc.

Mediterranean Breeds: **Ancona, Leghorn, Andalusian, Minorca, etc.**

Miscellaneous Breeds: **Hamburgh, Poland, Rumpless, Frizzled, Naked Neck, etc.**

Bantams: **Japanese, Pekin, Nankin, Belgian, Booted, Rosecomb, Sebright, Tuzo.**
From these came other bantams, miniatures of the large fowl.

KEY: SF = Soft Feathered
HF= Hard Feathered

The Breeds Structure

Breeds came from different lands and then were crossed. The Table is intended to illustrate what happened, but is not fully exhaustive.

From further East there was an influx of very large birds, but, oddly, were in a mixed type which varied from the very hard feather breeds such as the Malays and Sumatras, to the excessively feathered types, the principal being what became the Brahma and the other the Cochin. Another import was the Croad Langshan, a large majestic breed with greeny-black plumage.

Another important newcomer was the Aseel, the fighting, Game Fowl of India, which received special training in his own land. He was used extensively by W F Entwisle, the pioneer bantam breeder*, when developing new breeds in Victorian times.

These new breeds from China and India produced brown eggs and fresh blood from which was created many new breeds with distinct characteristics. Moreover, they introduced extra vigour which allowed heavier breeds to be developed as layers, better for winter laying because they were able to withstand the vigours of the climate.

Into the Melting Pot.

Inevitably, with the development of shows, which started around 1840, the first major event being the poultry show at the London Zoo in 1845, there was great interest in poultry breeds. The events received royal patronage as well as arousing the interest of notable people of the day. It is worth noticing that the gifted artist and author Harrison Weir was present at that very first major show and, after a lifetime of writing and drawing, he produced his great work

---

***For details of his work readers should see *Bantams*, available from the publishers.**

Our Poultry and All About Them. (1903), three years before he died at 82 years of age. This mammoth work provides invaluable information on the breeds.

From this beginning, with the extra blood lines at their disposal, and the high prices being paid for stock, it was inevitable that new breeds would be created. The Asian breeds in particular supplied the means of getting all kinds of permutations, which after stabilization and approval by the Poultry Clubs, here and abroad, they became standardized breeds. Many fell by the wayside and never really "took off", but others became well known breeds.

EXAMPLES OF BREEDS CREATED*

The new breeds brought a store of different sizes, shapes, and colours. Old English Game already possessed a range of colours, but the new imports widened the scope and created a vast range of different possibilities.

Faverolles (Salmon)
Light Brahma X Dorking X Houdan

Black Orpington
Minorca X Black Rock (Hen) X Langshan

Buff Orpington
Hamburgh X Dorking X Buff Cochin
OR
Buff Cochin X Dorking
which originally were known as **Lincolnshire Buffs.**

***Note: The formulae used is that recorded by early writers such as Lewis Wright and his contemporaries in the Victorian era.**

Original White Orpington with Rose Comb
Note similarity to White Dorking below and compare with Dorking page 88. White Dorking was an original colour and the White Orpington was produced in 1892.

White Orpingtons: the Evolution in stabilizing Breed and Colour.

Laced Wyandotte

Brahma X Spangled Hamburgh X Poland

Columbian Wyandotte

White X Barred Rocks

and then improved by crossing with the Light Brahma.

The Mixture

The inter-relationships which exist can be discovered by tracing the source of each breed and noting how the same breeds appear in each formula. An example should elucidate the point. For Black Orpingtons (above) the Black Plymouth Rock is one of the breeeds involved; this came as a sport from Barred Rocks which had been bred from Black Javas, Dominiques and the Cochin.

The careful intermingling of colours meant that very quickly the number of varieties multiplied. All was not straight forward because many breeders tried to produce in a season what should have taken a few years; with the enormous prices being obtained for new colours, these had to be produced as quickly as possible. The fact that such colours might not be produced again, apparently was of no concern to these entrepreneurs who were in it purely for gain. Yet from this confusing and often commercial approach there emerged the new breeds which were eventually to be standardized. There were of course the dedicated fanciers who were in the hobby for its own sake and it was to them that the development depended, and even the birds 'passed off' as something new and unique often did provide the nucleus of the new breed. There may be as many as 80 breeds available, although not all recognized in the UK or USA; the varieties make the number of colour variations much greater.

The time for perfecting a colour and, at the same time for it to be stable enough to be reproduced could take many years to stabilize. The experience of the author has been 4 or 5 years when introducing a new variety. Sir John Sebright took around 10 years to perfect the lacing on the Sebright bantam and many other cases have been recorded. In fact, for perfect colour reproduction from breeding stock a very long period may be involved; look at the improvements that took place over *decades* in producing the golden Buffs..

THE SCIENCE OF GENETICS

Poultry provide a complex structure of chromosomes and related genes and scientists have attempted to fathom their complexities with some success. With the development of microscopes and the methods of dating and analysing the cell structure (DNA), knowledge and understanding has multiplied.

Starting from basics it will be appreciated that the egg provides the means of breeding the next generation. When male and female mate the makeup of one meets the other and, through chemical change, when an egg is incubated, the new colours and other characteristics appear.

The instructions on what will develop come from the nucleus of the egg which contains the chromosome threads on which are contained the genes. These divide and then join back together so that the chicks contain the same number of chromosomes. It has been determined that man has 23 pairs of chromosomes and the domesticated fowl has 39 pairs, but this does not tell the full story because, whereas the fowl has up to 30,000 genes, humans may have in the region of 50,000 to 100,000. Moreover, the amount of DNA in the human genes is greater than in the domestic fowl. It has also been discovered that the idea of a chromosome being made up of beads (genes) on a thread is too simple an approach because the genes themselves are also broken down further into DNA units* as well as additional protein.

The diagram of the egg given should enable the reader to see what is involved.

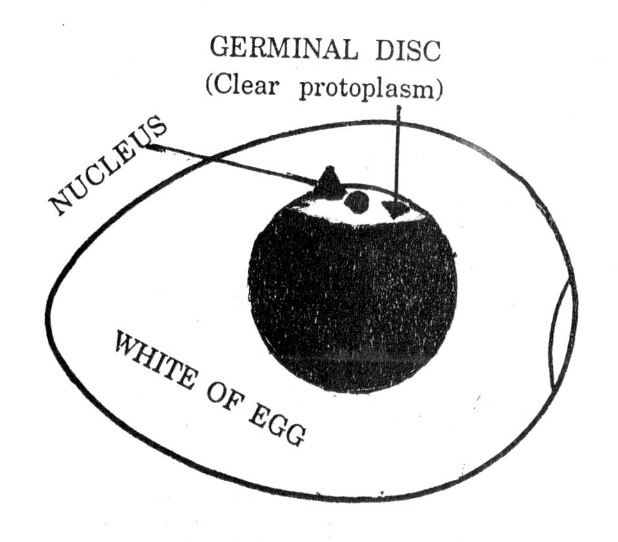

Beginning of the Life Cycle

*** This notion requires much study and explanation and readers are referred to Stevens (ibid)**

Various researchers have built "models" depicting the genes, and have compiled genetic maps, thus enabling a possible structure to be presented. However, whilst it is true that one or more genes determine colour, and there are others which decide patterns such as lacing or spangles, the complexity of the chromosomes make exact prediction very difficult. The general tendencies can be stated, which still follow the Mendelian theory, but how one gene may re-act on another can never be a foregone conclusion.

Note: An example of the scientific approach to breeding and the creation of new colours or a completely new strain may be seen from the breeding chart produced by a Mr Felch, an American breeder, who demonstrated that three distinct strains could be developed from one pair. This is explained later.

Jungle Fowl

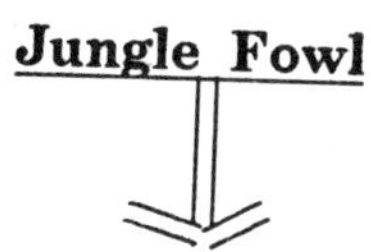

Old English Game

About 30 different colours + off colours

Dorkings

White
Dark
Silver Grey
Reds
Cuckoo

The Scots Dumpy ?

Jungle Fowl & the first Known Breeds in the UK

The exact path of the development is unknown, but we do know that many breeds were introduced, crossed with the above and further new breeds developed.

Breeds & Varieties

In the process of developing the different breeds, and it will be noticed here that the reference here is not to ***species,*** which is the description used for wild birds, but to the term used for the distinct shape or other characteristic of the different types. These breeds will cross with each other, denoting a common ancestor, usually reckoned to be one or more of the species of Jungle Fowl.

Many of the **breeds** are divided into different **varieties**, usually colour variations, some of which have a multitude of different forms. In fact, in some cases the colours have dictated to the form of the variety, so that in breeds like Wyandottes and Orpingtons, there is a difference in the basic types, because of the ancestor used to create the new colour. Thus for example, White Wyandottes conform to the 'bird of curves' description in exact terms, but other colours such as Buffs and Silvers tend to be lomger in leg and not so rounded.

There has always been a fascination for some colours and fanciers have gone to great lengths to create a particular colour in a breed. Different colours have come into fashion and have then been introduced to existing breeds. In the USA there seems to a fascination with this process. In the UK, Buffs and Blues were the favourite new colours, but in the USA different coloured tails have been a favourite introduction.

2

COLOUR CREATIONS

THE FIRST COLOURS

The general opinion seems to favour the view that all breeds of poultry originated from one or more of the species of Jungle Fowl, with the Red Jungle Fowl from Burma being the leading contender.* The colour of this species is Black Red for the male following the traditional pattern of black breast and tail and the remainder largely bright red. The female is a light partridge colour with a fawn to salmon breast, and an umber-yellow hackle. If we look at a typical Old English Game of the Black Red Partridge colour the combination is very apparent. This then should be regarded as the ***normal*** colour and, if so, other colours have been created from this combination by a freak occurrence or by strange behaviour in the genes pattern. In fact, it has been found that a specific gene has a number of possible behaviour patterns, or there are alternative genes on the chromosome,

*** There is a school of thought which argues that the large fowl of Asia, such as Brahmas, Cochins and Malays differ widely from the Red Jungle Fowl and therefore there must be another ancestor, possibly an extinct species. The *Gallus Giganteus* has been suggested. See *Keeping Jungle Fowl*, J Batty, where the full implications are covered.**

and these are known as ***Alleles.***

This combination of red, black and partridge is often referred to as the 'Wild' colour, denoting that this was the original colour. Once this is accepted, this becomes the standard pattern, and any variation can be assumed to be a departure from the wild type. With all such analyses it will be necessary to determine or be told which colours are **Dominant** and which are **Recessive**, because this fact will affect the results of any breeding in first and second generations.

The pigments which are of prime importance in determining the wild form; ie, Black Red, are Eumelanin and Phaeomelanin, If these are taken as flowing into a "receptacle" different colours will result, depending on the proportions of each at any given time. A dominant allele, which determines the colour is known as the **E *locus*** and the variations in this will determine the final colour.* Thus when birds are crossed there is no telling what might emerge, but with standard bred fowl of the same breed, the alleles are the same so the same colour can be expected. With others, from experience and previous knowledge, the expected result can be predetermined and good results obtained, but some will still be 'off-colours'.

This has been experienced numerous times by the author. Breeding Old English Game with the multitude of colours and 'off-colours' demonstrates very clearly what can occur with crosses within the same breed. Moreover, this approach does demonstrate the colour changes without affecting other characteristics, as would occur if crossing two different breeds. Even so, there are generally **'faults'** which

* **Lewis Stevens (ibid)**

Black Hennie Cock

Black Red

Golden Duckwing

Blue

Some Old English Game Colours

occur in the first one or two generations. Yet with careful selection, a fault, such as rust colour on the wings of a Golden Duckwing, can be eliminated and thus achieve an exhibition standard. Fanciers have been practising genetic breeding for generations, but without the background scientific knowledge; they could achieve an objective in colour breeding, but did not really understand the reasons. As a result of the selection which has occurred, OEG has more colours than any breed and, no doubt, many more could be achieved; examples are as follows:

Black Reds with Partridge, Wheaten and Clay hens.
Black-Breated Black Reds with Purple-black hens.
Black-Breasted Dark Reds with Dark Partridge hens
Brown Reds with Dark hens (Black/brown)
Gingers a self colour in both sexes, with dark/black colour in tail and possibly wings.
Blacks self colour, Blue bred
Blacks self colour, Melanistic bred.
Blues self colour, combined with black
Whites pure white both sexes.
Piles (Blood wing) as for Black Reds but white replaces Black.
Piles (Custard) Red hackles and shoulders replaced with yellow.
Blue Reds as for Black red Partridge, but Blue not black.
Lemon Blues as for Blue Reds, lemon replacing red.
Furnaces Red shoulders, rest black; hens black and brown mix.
Polecat Yellow shoulders, rest black, hens similar to Furnace.
Duckwing (silver) Black breast and tail, silver hackle & shoulders.
Duckwing (Golden) as for silver but deep yellow: hens for both a grey colour body.
Duckwing (Blue) Blue replaces the silver or gold.
Birchen Duckwing Brown shoulders replace yellow or silver as above.
Yellow Birchens a mixture of yellow and brown.
Spangles like the Black Red Partridge, but with spangled tips overall and burgundy coloured body in cock.
Creles and Cuckoos a form of barring.

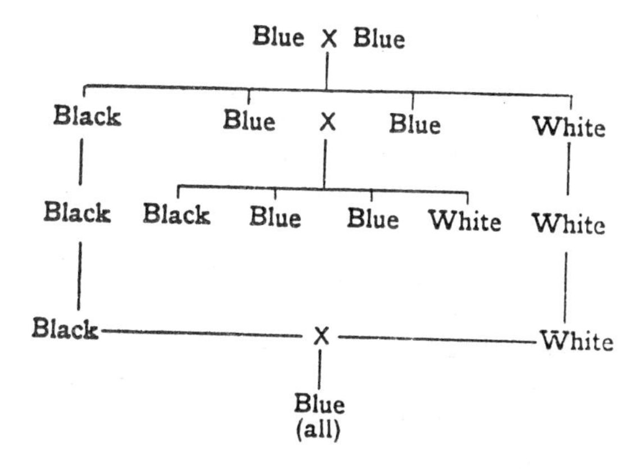

Unstable Blue (known as Andalusian Blue)

The difficulties with this colour are well known. As shown on the chart above, the Blue X Blue produce 1 Black, 2 Blues and 1 White in the first generation. The Black and White-Splashed bred together produce 100 per cent blues because, in effect, these are the 'pure' colours, whereas the Blue is not a true colour at all. (see *Mendelism*, R C Punnet, London, 1911.

This phenomenon occurs with other breeds so the principle is of great importance. However, there is now a stable blue, and, as a result, Blue can be bred from Blue; one fancier known to the author has produced Blue Leghorns from this colour, usually referred to as *Lavender*.

This assembly of colours gives the reader some idea of the store of colour genes available, and worked upon by Game fanciers over generations. These colours have also been used to create many of the colours found in other breeds. There are also many other variations; Blue-tailed Wheatens, Brown-breasted Brown Reds, Black-tailed Buff, BT Red, BT White (USA) and many others. In Old English Game bantams in the USA there are 29 different varieties listed.

One source of literature* has identified 69 different colours in poultry, which is not really surprising when the list of OEG colours are considered. The same author (Jeffrey) states that as many as eight alleles have been recognized on the **E *locus*** and there are others, particularly on the restriction of black.

THE LAWS OF COLOUR

Professor R C Punnet, following Gregor Mendel, Charles Darwin and others, found there were ***tendencies*** which usually occurred when breeding for colour. These are:

> **1. Whites may be different types; eg, may be Blue bred or may constitutionally be a "Cuckoo" type colour. It may be dominent or recessive; this may replace black, but does not always cover red, which is why Pile OEG, and similar colours exist.**
>
> **2.. Barring is really a self colour; eg, Black, to which a barring factor has been added.**
>
> **3. Blue is not a true colour and this is why 100% blue colour breeding is not possible. The true breeding forms are black**

*** *Bantam Standards*, American Bantam Association, quote in Bantam Breeding & Genetics, Fred Jeffrey, Hindhead, 1977.**

and white splash, and black. Note: In recent years Lavender Blue has been introduced into some breeds and this breeds true, but as yet, does not have a general application. It has certainly been introduced into Leghorn bantams.

4. Blacks can generally be produced from breeding black with black.

Note: This statement is only true in specific conditions because Blue-bred Blacks will produce Blues. There are different forms of Black, one with the blue content and the other with purple sheen or green sheen.

One of the greatest problems with blacks is where a bird has yellow legs, the production of the leg colour and the plumage colour clash and double mating has to be practised. This occurs very clearly with Wyandottes and Leghorns. Another diffficulty is the achievement of the correct eye colour, which may be too pale.

5. Variations in feather colour such as spangling, lacing and barring are due to a special factor in each case. The Spangle is a Black Red (wild genotype) to which is added a factor for mottling or spangles. The barring is caused by a sex linked gene which restricts the black colour, thus giving the white bands; the exactness of the bands appears to depend on the rate of maturity, so Plymouth Rocks which are slow to develop have more precise markings. A different barring is found on breeds such as Campines and Gold and Silver Pencilled Hamburghs. This results in narrower white bands and the gene is *incompletely* dominant.

From what has been stated it will be seen that the subject of colour breeding is quite complex and some knowledge of genetics is desirable. The section which follows indicates the main basic colours and tries to indicate their nature.

COLOUR BREEDING*

The original colour formations grew primarily as a camouflage for the birds in their wild environment. The surface colour provided the basic requirement, but in poultry breeding the *standard* often stipulates the colour of the ***under feathers*** (the fluff) so the pigment must extend right through to the skin.

Basic Colours

The basic colours found in poultry are:

Black
White
Blue
Buff
Red
Yellow

In fact,, when the primary colours used in printing are considered this number may be reduced to four: Blue, Red, Yellow, and Black, the rest being variations of these colours. With the addition of alleles which give variable patterns, there emerge all kinds of markings, such as spangles, lacing, mottles and splashes.

Blacks and Blues

The single colour bird is termed a 'self-colour' and this describes any bird with a single colour. However, in reality, if Black is taken, it will be found that this may be:

1. **Blue Black**
2. **Green Black**
3. **Purple Black**
4. **A combination of 2 and 3.**

***Dr W Clive Carefoot provided some of the information for this section**

White Langshan
(USA style)

White Faverolles

White Dorkings

White Orpingtons

Whites often Came Originally as 'Sports'

This colour has occurred in most major breeds.
Yet it is not dominant in many breeds.

The Blue-Black is well understood, the colour pigments being linked into a chain which included white, blue, black and a combination of black plus white, usually in irregular form, known as splashing. The blue in this type of colour is a dilution of Black. A Lavender factor reduces the black to blue; this also changes Red to Buff. If there is a second dose of this factor the offspring are black and white splashed, which explains the Mendelian theory.

A Black which does not seem to be understood is that contained on the plumage of the melanistic breeds such as the Crow-winged Blacks bred by our forefathers. Harrison Weir drew birds of this colour* and Herbert Atkinson mentions them++. The author bred them many times, but found that if the colour was diluted in any way by crossing with a lighter bird, the jet black with the purple sheen, was lost.

Turning back to Blacks generally, whether Blue-black or the Green-black of many breeds, it will be found to be a common colour; breeds include Jersey Giants, Sumatra Game, Hamburghs, Langshans, Leghorns, Minorcas, Orpingtons, Wyandottes and others.

Whites

True Whites may be one of two types:

Dominant (eg, Leghorn) *or*
Recessive

---

* In *Our Poultry*, Harrison Weir, then re-issued under the title *Harrison Weir's Game Fowl*, Editor, J Batty.

++ *Cockfighting & Game Fowl*, Herbert Atkinson.

Minorcas

Jersey Giants

Booted Bantams

Australorp

Blacks are an Interesting Colour

Appear in many different types so the correct form must be known.

The self White colour appears in many breeds such as Dorkings (with rose comb), Plymouth Rocks, Leghorns, Wyandottes, Rosecombs (Black is more normal), Orpingtons, Faverolles, Langshans, and Silkies.*

The dominant white is the result of a restricting factor which does not allow black pigment to form in the plumage. In some varieties it works on the black, but not on the red in the same way; thus Wheatens are produced and so are Piles. and Jubilee Indian Game.

Strangely, it seems the best way to a stable and pure white is to cross with a Black and this eliminates any yellow colouring. This is usually the case when the white is recessive; if it is dominant or dominant with recessive a sound white should be achieved.

Where the white is the result of breeding from coloured birds, this is a mutation, which generally means that the bird is unable to produce pigment in the feathers. Many heavy breed whites were produced in this way, being sports from another variety of the breed. It has been found that the sport bred back to the original birds will breed further whites, although this may be in the second generation.

***Silkies are a black skinned breed with mulberry face which have stimulated the interest of many researchers. For a detailed study of this unusual breed see *The Silkie Fowl*, J Batty, Elsted, 1996.**

Pile OEG Bantams

Jubilee Indian Game

Dominant White which produces some beautiful colours.

Pile and Jubilee Indian Game are produced from this gene being present.

Buffs

Buff is a popular colour which has been extended into many breeds. These include Brahmas, Cochins, Leghorns, Orpingtons, Plymouth Rocks, and Wyandottes. Originally there were serious problems in achieving a deep, even buff colour, free from light patches or black, although the latter is very difficult to eliminate altogether. The aim should be to get a golden colour within the buff, because this gives the depth required. Genetically it is quite complex, and is not fully understood. The undercolour must also be buff.

Because of the difficulty of eliminating black spots or marks from buff it has been suggested that Buff may really be part of a patterned plumage. Moreover, buff appears to be recessive to black. On the other hand, it is dominant to recessive white.

Black Reds and Their Variations

Black Reds are the original colours and come in many shades from very dark as, for instance, in Dark Reds in Old English Game to OEG Light Reds with wheaten hens. In addition, when dominant white is introduced a Pile type of bird is bred so that Black Red and Pile can be bred from the same pen; in fact, if Pile is to be kept to the Blood-wing variety an infusion of Black Red blood is essential every other year.

A form of Black Red also appear in Wyandottes (Partridge) and in Cochins (Partridge), Gold Brahmas, and Brown Leghorns. The colour gradations are different, but the basic colours are similar. However, as noted earlier, only the OEG Partridge hen has true Partridge markings.

OEG Black Red

Partridge Wyandotte

Partridge Cochin

Black Red Forms of Plumage

Only the OEG has the true wild colour.

Reds & Black-tailed Reds

Red colouring appears in many birds, but the shade can vary tremendously. Rhode Island Reds, strictly Black-tailed Reds, is no longer a normal red, but is a chocolate colour, including a strong infusion of black. At leasty three genes appear to be responsible for the deep red colour.

Reds like the ***Ginger Red*** in OEG are a more true red with a dark tail. The author has found that these will inter-breed with Brown Reds to produce a darker bird, although this can be bred back to the brighter ginger by selective breeding.

Some varieties are called 'Red' when, in fact, they are Black Red; eg, Red Dorkings. The New Hampshire *Red* is a from of dark Buff which has a black ticking in the female's hackle. The Red has the gene for brown, whereas the NHR has the gene for Wheaten, as well as gold and others.

Silhouette of Rhode Island Reds showing 'Brick Shaped' Body.

Must be a chocolate colour to win (Colour Plate)

FURTHER VARIATIONS

There are many variations found in practice from the Sebrights with their beautiful lacing, the Lakenvelder with black tail and hackles, and many other colour patterns.

Often, as shown, a gene may be one which ***restricts*** rather than does something positive. With other cases, the impact of one gene on another can have far reaching effects on the ultimate colour. Fortunately, once a breed is established the birds breed true, except where the *standard* is out of line with practical reality, when double mating may be required; in effect, rather like producing two separate breeds.

Double Mating

When the standards were originaly compiled, in those far off days when poultry showing started, the fanciers were often concerned with what they would like to see, rather than what was a reality. In addition, there were artists like Ludlow and Wippell who tried to show pictorially the ideal birds as envisaged by the standards. In some breeds they were well ahead of their time; in others they were showing pairs of a breed which could never realistically be regarded as an ideal breeding pair, which would produce the same as their parents. The genes were 'incompatible' and therefore the offspring would not be in accordance with the standard.

The depth of colour required for say a Black cock would not be compatible with the colour specified for the hen. If Dark X Dark matings were used the result might be all very dark offspring, but if the standard required the male to be dark and the hen medium, there was no way exhibition type hens would be produced. This had to be done through separate breeding pens, and thus was ***double mating*** created.

Sebright Lacing

Without correct lacing the breed has no chance of winning.

Double Lacing **This appears in Indian Game (hen only) & Barnevelders (Male & female)**

Single & Double Lacing

Fanciers began to realize that key features (those which won prizes) required special attention. For some breeds they had to be bred in different pens, using in some cases ***cock-*** *(or hen)* ***breeding females*** or ***hen*** *(or cock)* ***breeding males.***

Examples of these special features are as follows:

1. Depth of Colour - Blues, buffs and Blacks.

2. Barring.

3. Lacing which is clear and even.

4. Elimination of colour faults such as unwanted ticking, lacing and incorrect shading; also sootiness, peppering and other mismarking.

5. Hackles not "matching" in male and female.

Birds must be mated which produce the desired features in:

(a) Male, and, *separately,*

(b) Female

This means that it will be necessary to determine the **Ideals** for:

1. Prize winning cock for breeding with a female with male colour tendencies.

2. Prizewinning hen to breed with a cock with female colour tendencies.

Not all fanciers agree with this approach and would prefer to see the ***standards*** revised to allow males and females of the desired quality to be bred from the same pen.

INTRODUCING A NEW COLOUR

The poultry fancy is full of examples of new variety creations and the spreading of new colours. Such colours as Blues and Buffs have become fashionable and have been introduced to most multi-variety breeds. Barred and Columbian patterns have also had their share of popularity.

Broadly speaking colours can be introduced in one of two ways (or a combination):

1. From a chance breeding, usually referred to as a "Sport". This can occur at any time, but the most likely is when two 'lines' are crossed, thus playing havoc with the mixture of genes and throwing out many possibilities.

Note: This very morning as I write I have an Old English Game pullet who was sitting on 8 eggs and has today produced 7 chicks. There is certainly a Blue, possibly two Blacks, three partridge (with the wild fowl twin, light stripes down the back) and the other, possibly, a Duckwing. These were bred from Black Red/Partridge, but last year I crossed the strain with a very shapely Lemon Blue cock and the pullet was from that cross, but ostensibly a Partridge colour.

2. From the correct colour on one side, matched to the breed required to be made into a new variety.

Entwisle* explains how he first introduced the Cuckoo Pekin. From a White Booted X Black Pekin he obtained one offspring pullet, which had faint barring. He crossed this with a Black Pekin cock and produced further, but more distinctly marked pullets. These were then inbred and further selected until the new variety was able to win a cup at Crystal Palace.

The process of perfection can take many years, and may be very difficult if the breed introducing the colour has undesirable features. Thus to take an extreme example, do not attempt to introduce a new colour from Silkies which are quite different from all other breeds; eg, silkie feathers/ black skin.

---

* *Bantams*, W F Entwisle, available from the publishers.

The possible reasons for the long period involved in stabilizing a colour are as follows:

1. In the first crossing of a male and female (known as the ***First filial*** or **F1 generation**), the results will depend upon which genes are dominant and which are recessive.
2. When the next cross is made within the same group, the results will be different because now the genes are combining in a different way.
3. When unrelated birds are first crossed there is a tendency for ***reversion*** to the chromosomes of an early ancestor.

EXAMPLES

Black Orpington X Black Orpington

both from same strain

Expectation, say, 95 per cent of soundly coloured birds.

Black Orpington X Light Sussex

results would be unpredictable because now the genes are mixed together in an unknown way.

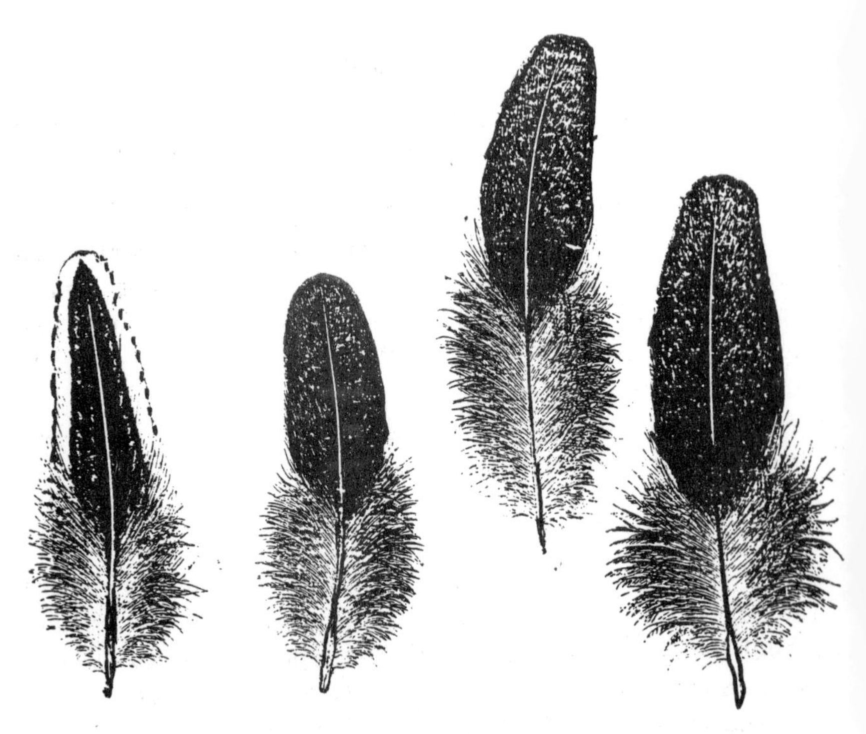

Hackle	**Flat of Wing**	**Back**	**Saddle**
Golden Streaked with Black	**Body Partridge which is a mixture of brown and black; breast salmon.**		

Examples of Black Red/Partridge HEN Feathers
Found in Old English Game and Modern Game

3

COLOUR PATTERNS

BREEDS WITH CONFUSING DESCRIPTIONS

The poultry fancy has grown haphazardly, and breed clubs have considered their own varieties, but not always considering what had already taken place, including the terminology adopted. The result is that terms used are not readily understood and may even be misleading.

The purpose of this chapter is to examine and explain some of these terms or misconceptions.

Black Red

This is the original ***wild-colour form*** and consists of a male with black breast and tail, with red hackles and shoulders. The female is a partridge type with a brown body with markings, a fawn hackle with darker stripes and a salmon breast. It is best known in Old English Game.

It also appears in many other breeds with the same name, as for instance, Malays and Aseels.

With some of the soft feather breeds other descriptions are used. Thus in the following:

1. Brown

This applies in Leghorns and in Sussex.

Grey Old English Game

Modern Game (Birchen)

Grey & Birchen - Terms which are confusing

In some other breeds are given different names; eg, Silver Sussex.

2. Partridge

Cochin, Plymouth Rock, and Wyandotte. Other are Partridge–types, but are not usually referred to with that description. This is also the case with Welsummers.
As pointed out on a previous occasion* the only true partridge is found in OEG where the hen is peppered all over in black and there are no mating problems with colour. See illustration of feathers on page 34.

Grey and Birchen

Confusion reigns supreme with the term 'Birchen'. The description is used to describe a yellow cock with brown breast (OEG Birchen Yellow) which is correct because birchen is a brownish colour. It is also used for Modern Game to describe a black bird with a grey striped hackle and silver grey shoulders, known as a Grey for Old English Game. Jeffrey (ibid), wrongly, appears to believe that a Birchen is a Grey with less body lacing.

This appears to be one of those cases where usage over time, especially with Modern Game, has established an incorrect name. Efforts should be made to call a Grey by its proper name.

"Lights" and Columbian Patterns

There are a number of breeds which are virtually self-coloured, but have a glossy black tail and a hackle with

---

* ***Why Double Mate,*** **W H Silk, Chairman of Standards Committee, Poultry Club, *Poultry Club Yearbook,* 1950.**

a narrow lacing of silvery white. This pattern of colouring may be found in, Columbian Leghorns, Columbian Wyandottes, Columbian Plymouth Rocks, and is also in *Light* Sussex and *Light* Brahmas.

Because the colour pattern appears in many breeds the type should be observed so it is understood. The body colour should overall be quite clear and unblemished - if white it should be 'of the purest white'. Buffs should be an even golden buff. Originally it seems that the term was restricted to White birds with black markings, but now embraces buff and others are posible.

In the conventional Columbian there is a gene which restricts the black except in certain parts; ie, hackle, tail and wing feathers.

The White males should have a silvery white hackle marked with an intensely coloured stripe in black, but with a clear margin of white all round the black stripe and without any black tip at the base.

The tail should be green, glossy black. The tail coverts should be black laced with white.

The wings should be black and white following a well defined pattern. When opened it will be seen that the primaries are black with lower edge white. The secondaries are divided into black (unexposed) and white (exposed portion).

Columbians are auto-sexing breeds. For breeding stock to produce the medium black required it is usual to select birds which are dark and light, thus getting the balance required.

This colour is related to the Lakenvelder and to the laced breeds, such as Laced Wyandottes, as well as Campines and some of the Belgian bantams (Millefleur a ıd Porcelaine).

Columbian Plumage
Columbian Wyandotte

Columbian-Type Feathering
Courtesy: Irvin Holmes

Lakenvelder Feathering

Comparing the Patterns

Laced and Pencilled Varieties

In some breeds the description is quite adequate, but in others is not correctly described. Lacing should be a border of a different colour around the edge of a feather. It may be ***single***, eg, Sebright, or ***double,*** eg, Indian Game or Barnev-elder.

Pencilling is fine lacing or bands which appear in "rings" around each feather in a uniform way; eg, on the hens of Dark Bramas, Silver Grey Dorkings, and Partridge Wyan-dottes.

On Pencilled Hamburghs, although called 'Pencilled',in fact, are a form of barring, which causes some confusion amongst fanciers. The differences are covered more fully with the descriptions of certain breeds, such as Hamburghs and Wyandottes. The marks referred to as 'Pencilled', but in re-ality are bars, are shown below as well as on the ***Types of Feathers*** indicated overleaf.

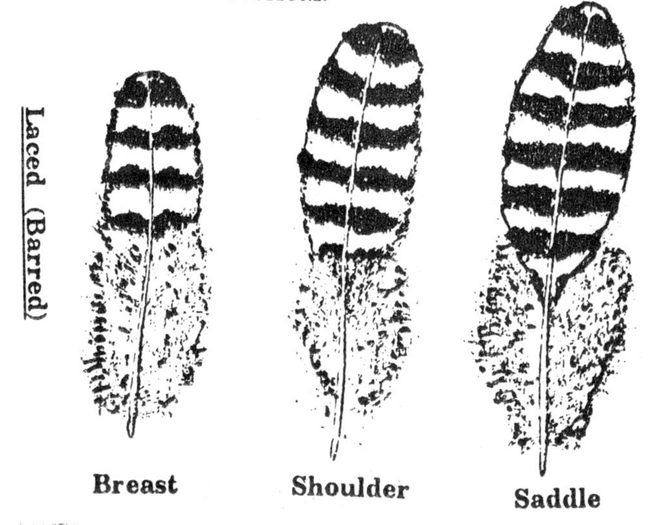

Pencilled Hamburgh (Female) Lacing

Barred and other Types of Marked Breeds

There are many breeds which have bars running across the body and tail. In conventional barring the bars are approximately of equal width, being a black or blue bar, alternating with a whitish colour. The barring gene is an inhibitor of the main colour such as black, thus giving the white barring. As noted, the barring gene is incompletely dominant; it is also sex linked so the chicks can be distinguished at birth by the cockerels being black with a white head spot and the females being self coloured black. This would occur when crossing a barred bird with a non-barred; eg, Black Langshan cock X Barred Rock females or Rhode Island Red male X Barred Rock when the females are not barred but the cockerels are. This occurs by crossing the Dominant female with a recessive male, the gene being sex linked for both.*

The breeds affected by the normal barring gene are: **Dominques, Leghorns (Barred and Cuckoo), Malines, Marans, North Holland Blue, Old English Game, Pekins, Plymouth Rocks, Scots Greys and Scots Dumpies. Also sex linked breeds such as Brockbar, Rhodebar, Gold Cambar, etc. The so called Creles, Mackerels and Cuckoos (OEG) also come into the barring process.**

The degree of barring or the strength of it will be affected by the genes within the breed and the time over which the selection for barring has taken place. In the Barred Plymouth Rock the barring is of a very high standard, whereas in the Marans the bars are indistinct and are referred to as 'Cuckoo'. Some breeders have suggested that the time it takes

*** For a fuller explanation the reader is referred to Stevens (*ibid*) or Jeffreys (*ibid*).**

Plymouth Rock — Scots Grey

Malines — Dominique

Typical Barred Breeds

1. Golden Pencilled Hamburgh Cock's Sickle
The outer margin on the black sickle is a golden colour. Line on outside simply indicates outer margin.
2. & 3. G. Pencilled Hen Feathers.
4. Partridge Cochin.
5. Silver Laced (Wyandotte Cock's Hackle)
6. Silver Laced Breast (Wyandotte Hen)
7. Silkie (No webs)
8. Double Laced (Indian Game Hen).
9. Spotted (Guinea Fowl)
10. Barred Plymouth Rock Hen
11. Plymouth Rock Male's Hackle
12. White Spangle (Houdan)
13. White Spangle (Houdan Male's Hackle)
14. Black Spangle (Silver Spangle Hamburgh Sickle)
15. Black Spangle (Silver Spangle Hamburgh Hen's Breast)
16. Black Spangle (Silver Spangle Hamburgh Cock's Hackle)

Source: *Encyclaepedia of Poultry*
J T Brown

KEY: Types of Feathers (opposite)

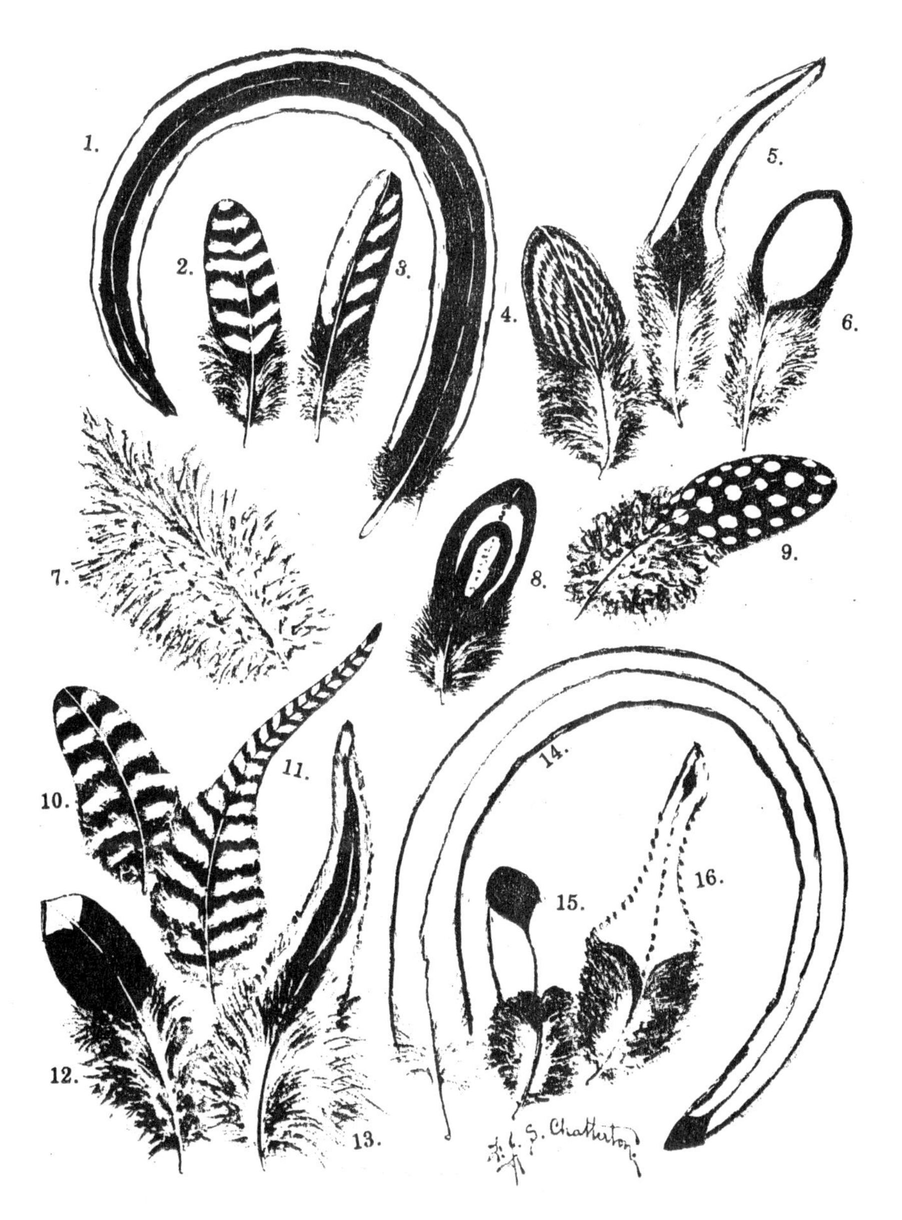
1.
2.
3.
4.
5.
6.
7.
8.
9.
10.
11.
12.
13.
14.
15.
16.
F. J. S. Chatterton

for a breed to mature is an important factor, but there seems to be no proof of this supposition. However, there is no doubt that the cuckoo-feathered breeds do feather more rapidly than the barred with their precise markings. The persistent call for improvement by judges over a long period seems to have been the main determinant.

A further distinction between Barred and Cuckoo is base on the shape of the bars and their colour. Cuckoo bars are usually regarded as crescentic in shape and greyish-white in colour.

In addition, because of the *standard* requiring clear, bold barring, not too pale, double mating has been practised, along with very strict selection of the birds to keep for exhibition. Barred to Barred produces varying levels of colour, some dark, some medium, and others light. **Barred female X Black male** will produce a mixture of each, and will be sexed linked. Lightly barred birds together should not be used for breeding because usually they only produce light colours; however, Light barred male X Black female should give good results because all the chicks should be barred.

Another form of barring is to be found with Campines, Pencilled Hamburghs (Note: the ***pencilling*** is really barring) and various Partridge varieties. The gene in question is what is known as an **autosomal barring gene,** usually designated by the letters Ab, whereas the normal (Incompletely Dominant) barring gene is B, BB or Bb, depending upon whether of single or double strength.

Related to the barring genes is one which produces the white tips on breeds like the Houdan, Ancona and probably the Spangled Sussex, which is not a true spangle at all. This is an **autosomal recessive gene** which gives mottling.

Spangled is related and the mottling gene (Mo) added to a dark Partridge-bred Black Red will produce the Spangles.

Although it seems certain that the Speckled Sussex is related to the Spangle OEG, in the sense that this probably is where the spangles came from, the Speckled does have an additional gene which produces the dark burgundy. The suggestion made by Fred Jeffrey is the presence of Columbian, but he does not explain further. Certainly the gene for the Mahogany colour appears to be in the make-up, which he also acknowledges. Old English Game bantams have the burgundy colour on the male.

Silhouette OEG Spangle Bantam

This variety must have even spangling - not as important with Speckled Sussex.

COLOURED ILLUSTRATIONS

Plate 1

Ancona - Black Mottled
Andalusian - Blue
Araucana - Lavender (Many Colours)
Aseel - Spangle (Many Colours)
Australorp - Green Black
Barnevelders - Double Laced (Number of Colours)

Plate 2

Brahma - Dark (Number of Colours)
Belgian Bantams - Millefleur (Many Colours)
Campine - Silver (Two Colours)
Cochins - Partridge (Many Colours)
Croad Langshan - Black (Two Colours, but black usual)
Dorking - Dark (Many Colours)

Plate 3

Faverolles - Salmon (Number of Colours)
Hamburgh - Gold Pencilled (Two Colours; two sub-varieties plus Black in Large)

Houdan - Black Mottled
Indian Game - Dark (Three Colours)
Leghorns - Cuckoo (Many Colours)
Nankin Bantams - Nankin colour Dark Ochre and Yellow

Plate 4

Malines - Cuckoo
Modern Langshan - Black, but White known
Marsh Daisy - Wheaten or Golden (Number of Colours)
Old English Pheasant Fowl
Orloff - Spangle (Number of Colours)
Orpington - Blue (Many Colours)

Ancona
Andalusian
Araucana
(Lavender)
Aseel
(Spangle)
Australorp
Barnevelder
(Double Laced)
Charles Francis

Brahma
(Dark)
Belgian Bantams
(Barbu d'Uccles: Millefleur)
Campine
(Silver)
Cochins (Partridge)
For British Bantams see Pekins
Croad Langshan
Dorking
(Dark)

Faverolles
(Salmon)
Hamburgh
(Gold Pencilled)
Houdan
Indian Game
(Dark)
Leghorn
(Cuckoo)
Nankin Bantams

Malines
(Cuckoo)
Modern
Langshan
Old English Pheasant Fowl
Marsh Daisy
Charles Francis
Orloff
(Spangled)
Orpington
(Blue)

Marans
(Dark Cuckoo)
Naked Necks
(Transylvanian)
New Hampshire Red
North Holland Blue
Charles Francis
La Fleche
Malay
(Black Red)

Plymouth Rock
(Barred)
Polands
(Gold)
Redcaps
Rhode Island Red
Charles Francis
Scots Greys
Sussex
(Speckled)

Silkie
(Blue)
Welsummer
Yokohama
(Black Red)
Wyandotte
(Gold Laced)
Modern Game
(Pile Cock & Hen;
Partridge Hen)
Old English Game
(Large Brown Reds;
USA Bantams)

Rumpless Bantam
Pekin Bantams
(See also Cochins)
Frizzle Bantam
(Blue)
Japanese Bantams
(Greys)
Old English Game Bantams
Gold Sebrights
(Black Red: English type;
See large Game for USA)

COLOURED PLATES (Cont)

Plate 5

Marans - Cuckoo (Number of varieties in Cuckoo)
Naked Necks (Number of Colours)
New Hampshire Red - Orangy Red with Black hackle marks
North Holland Blue - Blue Cuckoo
La Fleche - Black
Malays - Black Red/Clay (Many Colours)

Plate 6

Plymouth Rock - Barred (Many Colours)
Polands - Gold (Many Colours: two sub-varieties)
Redcaps
Rhode Island Red - Chocolate Red (White also known)
Scots Grey- Cuckoo
Sussex - Speckled (Many Colours)

Plate 7

Silkies - Blue (Number of Colours)
Welsummer - Black-Red type (Also Silver Duckwing in large fowl)
Yokohama - Black Red ((Number of Colours)
Wyandotte - Gold Laced (Many Colours)
Modern Game - Pile Pair & Partridge Hen (Many Colours)
Old English Game - Brown Red (Many Colours)

Plate 8 (Bantams Only)

Rumpless Bantams Golden Duckwing (Number of Colours)
Pekin Bantams - Black (Many Colours)
Frizzle Bantams - Blue (Number of Colours)
Japanese Bantams - Greys (Many Colours)
Old English Game Bantams Black Red/Partridge (Many Colours)
Sebrights - Gold (Two Colours - Gold & Silver Laced)

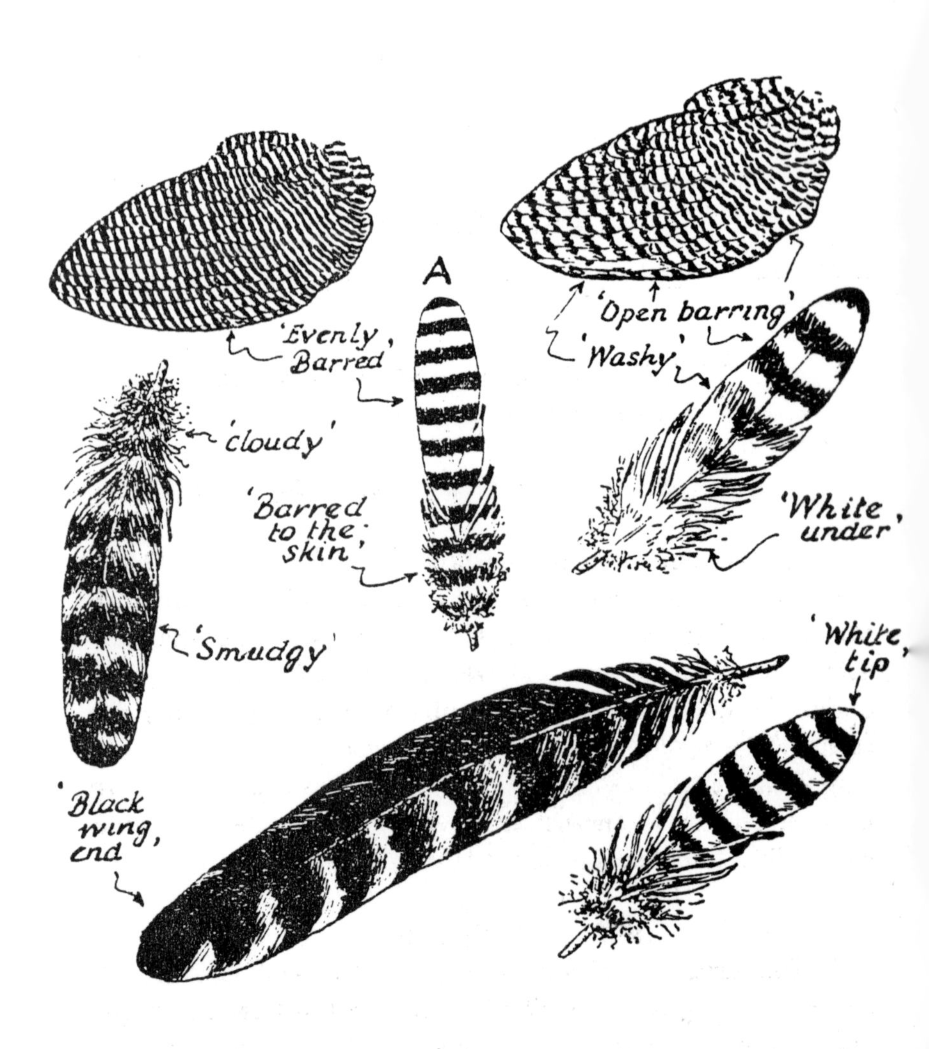

Features & Faults In Barring (Plymouth Rocks)

4

THE BREEDS

POULTRY BREEDS AND VARIETIES

There is an adage that states:

Type makes the breed and colour the variety

In the main this is true because type relates to shape, carriage and other physical characteristics which distinguish the breed. The colour provides the differences and therefore establishes the variety, and whilst some other feature may decide the different breed; eg, muffs or tassels, generally the colour variation is the determining factor.

Some breeds only have one colour whereas others have a colour similar to another so, by cross referencing, it should be possible to determine the colour applicable.

CLASSIFICATION

The first breakdown is into LARGE and BANTAMS, the latter being 25 per cent of the size of the large. In addition, there are natural bantams which do not have an equivalent large breed.

Then there is a further classification into **groups** as

shown:

1. Light Breeds such as Anconas, Leghorns, Old English Game which are around 6lb. in weight (2.70k) and proportionate for bantams. Usually these are non-sitters, although some, like OEG, do come broody.
2. Heavy Breeds which include Brahmas, Cochins, Dorkings, Modern Game, Indian Game and Wyandottes. Usually these are sitters; they weigh from around 2.70 kilos upwards to 5 kilos or beyond.

This division is rather arbitrary because some breeds are fairly heavy; eg, Spanish which are around 7lb in weight and a light breed, whereas the standard Silkie is no more than 4lb.

Another grouping is into:
(a) Soft Feathered, which includes those birds which have a profusion of feathers.
(b) Hard Feathered, which cover all the Game breeds, such as OEG, Indian Game, Sumatra Game, Malays, Shamos, Tuzos, and Aseel.

Some breeds are ***medium feathered*** and include Hamburghs, Scots Greys, and some of the Mediterranean breeds so a medium category would be appropriate.

All are described, and judged on, the book of *standards* issued by the poultry clubs in the appropriate country. The fancier must be prepared to read the appropriate ***standards*** and know its contents so that the type and colouring being bred are correct.

Comparison of Heads
The cock on the right is the ideal

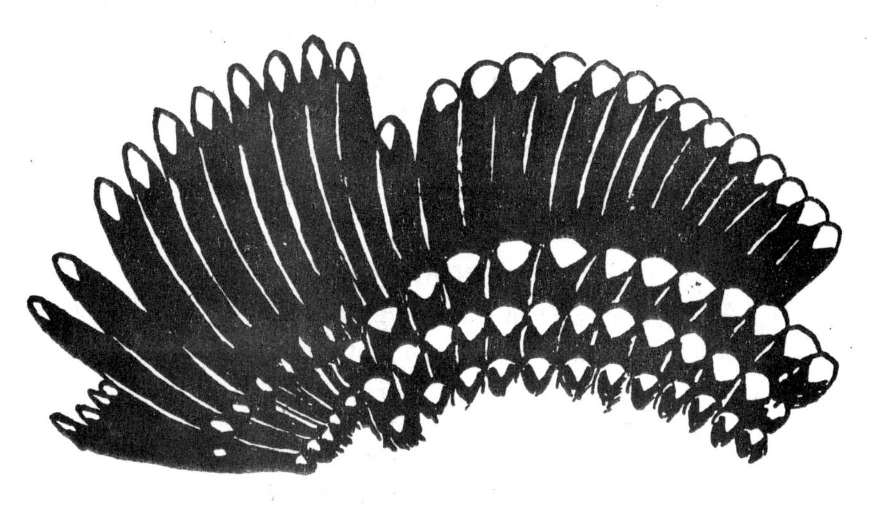

The Wing with the correct tipping of all feathers.

Ancona Mottling or 'Tipping'.

ANCONA*

This Italian breed exists in one colour only, although oddly there are two varieties based on a single comb or a rose comb. It is an excellent layer with agile, active movements, making it one of the workers of the poultry world.

Main Colour: Beetle green plumage which is a black with a green sheen; each feathers is tipped with white V-shaped marks in regular sequence. Lack of tips, too many, or the wrong shape, or too large, are all faults. Legs yellow with black mottles and eyes reddish.

Its main drawback is that it tends to be rather highly strung so is not suitable for the back garden, except in bantam form.

ANDALUSIAN*

This is a fowl famous for its blue colour, which should be a nice, even blue; although slate colour with darker lacing – the overall colour should not look navy blue.

Faults in colour include white, dark patches, unevenness, sooty background, or the appearance of red, yellow or other colours. The ear lobes should be white and the eyes reddish. Legs are dark.

ARAUCANA*

This breed originates from Chile and may be Rumpless (no tail) or with a full tail. Its unique feature is the fact that it lays blue/green eggs.

Main Colours: The colour illustrated is Lavender, which is the most popular but a multitude of colours exist along the lines of OEG – Black Red, Brown Red, Golden Duckwing, Cuckoo, Pile, Black and White, and Greys. **Faults** are any marked deviations from the main colours of OEG.

*** See Plate, page 48, for example of main colour or one of colours.**

Spanish
Black plumage with emphasis on ornamental face

Leghorn
Black

Minorca

Andalusian

The Light Breeds

These are mainly Blacks, or Blues, but in Leghorns many colours developed.

ASEEL* (Asil)

The colour illustrated is the Spangle which is quite popular. **Other Colours**: Dark and Light Black Reds, Blacks, Whites, Blues, Yellows, Greys, Duckwings, and Grouse. Although it is suggested that colours should follow OEG in reality the colours are not so distinctive, neither need they be, for colour is not of great importance in such a minority breed. Eyes should be a very light pearl or light yellow and legs to match plumage colour. These are very aggressive birds being trained for fighting in India for hundreds of years.

AUSTRALORP*

Colour is Black with a brilliant green sheen. This was the original Orpington which went to Australia as that breed and when returned was no longer like the show Orpington which had become very feathery. Eyes, beak and legs should all be dark.

Faults in colour are any tinges or marks in the green black

BARNEVELDER*

Main Colour: Double Laced requires both male and female to be reddish brown with black, double lacing, which is illustrated. The remaining colours, rarely seen, are Black, Partridge, and Silver. **Fault** is white in coloured plumage.

Black should be an overall colour with a beetle green sheen.

Partridge is the Black-Red type with the female showing more brown than the Double Laced variety.

Silver is not found in bantams, but in the large fowl the cock is a silver colour with black lacing and black in tail and wing primaries. The female is of a similar colour with black hackle with white centres.

Legs yellow in all varieties and eyes orange.

*** See Plate, page 48, for example of main colour or one of colours.**

Appenzeller Bantam

Bresse - Black & White

Some of the Minor Varieties

OTHER BREEDS - NOT SHOWN IN COLOUR

Appenzeller

A Swiss breed with a crest which bears resmblance to a 'crew cut' in hair styles.

Main Colours Appenzeller Spitzhauber:
Black, Golden Spangled, and Silver Spangled.
V-shaped comb.

Main Colours Appenzeller Barthuhner:
Partridge, Black and Blue.
Rose combed.

Booted Bantam

Also known as Sabelpoot is a very close relative to the Belgian Bantams; in fact, treated as such in USA. However, there are differences and it should not have the 'bull-neck' of the Belgians. It is feather legged and vulture hocks appear to be acceptable.

Main Colours:
White, Spangles, Black, Black Mottled, Millefleur and Porcelaine. The descriptions may be found under Belgians.

Bresse

Rather like a heavier version of the Leghorn.

Main Colours:
Black, and White.

Catalina

Bantam of Mediterranean stamp.
Main Colour: Buff.

Creepers
These are similar to Scots Dumpies.

Creve Cocur

Some Unusual Breeds

Chantecler

Rather like an Indian or Cornish Game in type. Canadian breed.

Main Colours:

White, Partridge.

Cornish

Based on the British Indian Game, being broad and heavy with deep yellow legs.

Main Colours: Black, Blue, Blue Laced Red, Buff, Columbian, Dark, Mottled, Spangled, White, White Laced Red.

Many of these colours are only in bantams.

Creeper

A general term given to breeds which have very short legs and bred together have lethal genes. Scots Dumpies, German Creepers, Danish Creepers, and Courtes-Pattes come under this general heading. The Japanese bantam is also a form of Creeper.

Main Colours USA Bantam Creeper:

Black, White Silver and Cuckoo

For different breed colours see under appropriate entry.

Creve Coeur

An unusual French breed with a V-shaped comb and crest.

Main Colour:

Black with a green sheen.

Columbian Wyandottes (Top) & Light Brahmas

These follow the traditional Columbian-type plumage and are very similar in type. Differences are in comb and feathering of legs.

BRAHMA*

Main Colours: Dark, Light, Gold (Partridge), Buff and White. The legs are feathered, but vulture hocks should be avoided (see Sultan for example).

Dark (also known as Silver Pencilled) a combination of black and silver in the male and grey with pencilling in the female (see Pencilling Description earlier.)

Light (also known as 'Columbian') which is primarily a white bird with black markings on hackle and tail. **Buff** (Columbian) is similar, with Buff golden buff plumage.

Gold or Partridge: a rich gold colour in male with black breast and tail. The female is a gold partridge colour (see Partridge)

White is white overall.

Legs should be yellow and eyes reddish bay. **Faults** are any colours not in accordance with colour standard.

BELGIAN BANTAMS*

These are ***true bantams*** with no large equivalent. They are exotic and ornamental and in two sub-breeds (note the difference in terminology - see ***Type makes the Breed*** earlier). They are known as:

1. Bearded Antwerp, and 2. **Bearded Uccle,** Barbu d'Anvers and Barbu d'Uccle respectively.

The first type is rose combed and clean legged, whereas the second is single combed and has feathered legs.

Main Colours are:

(a) **Millefleur** Orange red with mahoghany red on the wing bows of the male. Overall an intricate pattern of black spots and white triangles.

(b) **Porcelaine** A beautiful creamy colour background ("Light Straw") with the pattern of spots in blue with white triangles.

* **See Plate, page 48, for example of main colour or one of colours.**

The Modern Langshan
USA type tends towards the Modern

(c) Self Blues (There are many variations in Blue)

As for other blue breeds in a pale shade. An Andalusian Blue.

(d) Cuckoo

Light grey with bars of a darker shade.

(e) Quail

A mixture of black, gold and umber with the black predominating on hackle and tail.

(f) Black Mottled

Green black with an even distribution of white tips on the feathers.

CAMPINES*

A light weight breed with the male 'hen feathered', although many cocks do have a slight curve in the sickle feathers.

Main Colours: Gold, and Silver.

Both are **barred (autosomal barring)** with black (beetle green), with ***precise*** bars three times the width of the gold or silver ground colour, and extending across the tail. The hackle is silver or gold, depending on the variety.

The colour of the **legs** is leaden blue. The **eyes** should be dark with a distinct pupil. Faults: any departure from preciseness of bars with two colours intermingling.

COCHINS*

This very large breed with feathered legs and single comb is quite majestic, and similar to the Brahma (rose comb).

Main colours: (Note in USA which has Cochin bantams 16 are listed, equivalent to British Pekins).

Black: Self; **Blue:** Pigeon Blue; **Partridge:** really a type of Black Red (see illustration); **Cuckoo:** Blue/grey pencilling or bars; **Buff:** Deep golden buff without fading or patches; **White**: Self.

*** See Plate, page 48, for example of main colour or one of colours.**

Silver Grey

Dorkings

Oldest English soft feathered breed

CROAD LANGSHAN* (USA: LANGSHAN)

These were brought from China by a Major F T Croad, and were extended into Modern Langshans by the 'modern', long-legged craze which swept the Victorian show world.

Main Colour: Black with a brilliant green sheen. In the USA there is also a White variety.

Legs in Black should be dark with scanty feathering; eyes brown.

Faults: Purple in excess, other colours in plumage. Wrong colour legs, yellow or white.

DORKING*

The Dorking is an ancient breed famed as a table bird. The comb may be single or rose depending on variety. It has white legs and skin and has five toes.

Main Colours:

Silver Grey: Male has silver hackles and shoulders, and black breast; the female is a lightish grey colour with partridge markings.

Dark (also **'Coloured'**): A darker version of the light grey; in fact, the light grey was bred from the Dark (see colour illustration). In fact, the present Dark has been made lighter by crossing back to the Light Grey.+

Red: These are really a type of Black Reds with black breast and tail and the shoulders a deep red; the hen is brick colour in body with a sort of lacing on each feather, and an orange hackle with dark stripes.

White: Self with rose comb.

Cuckoo: Typical cuckoo, bands of dark grey or bluish grey, alternating with lighter colour.

*** See Plate, page 48, for example of main colour or one of colours.**

+ For original colours see *The Sussex & Dorking Fowls*, J Batty.

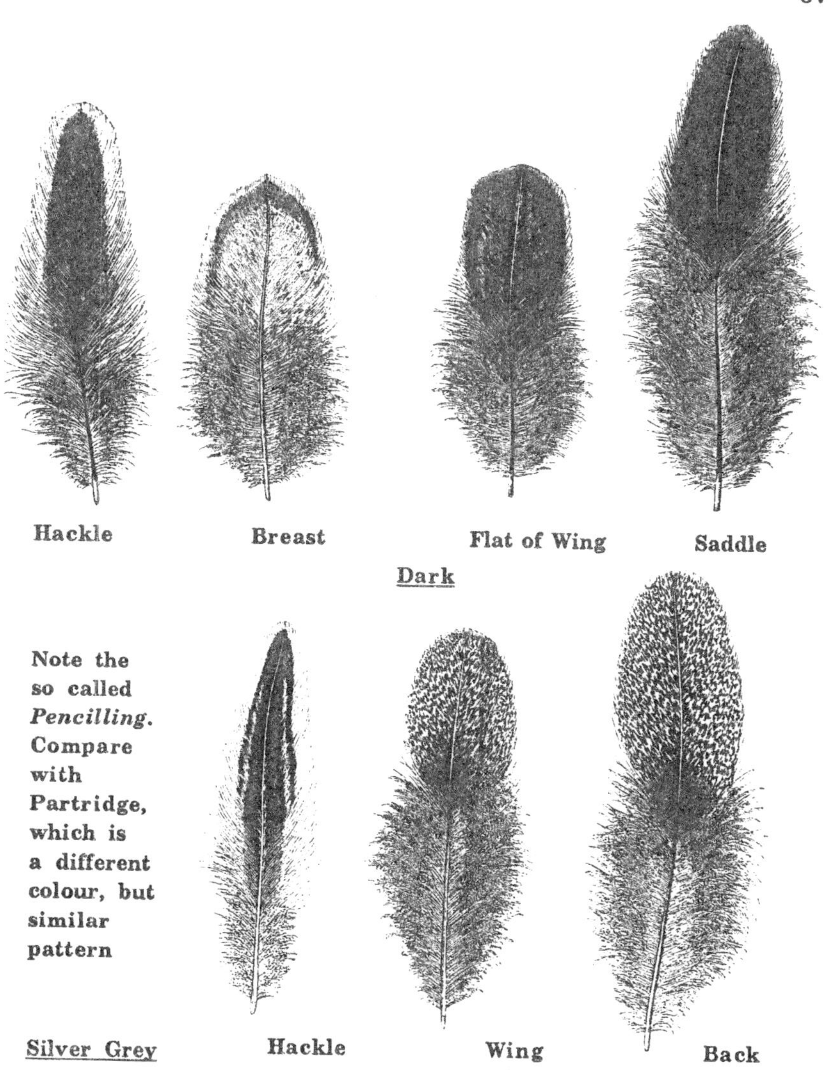

Feathers of Dorking Hens

Great care is required to breed the proper colours and any indiscriminate crossing of colours can spoil the end result; eg, blurred colours, rust on wings of Light Greys.

Dutch Bantams
Colours follow stamp of Old English Game, although not as numerous.

OTHER BREEDS - NOT SHOWN IN COLOUR

Cubalaya

Long-tailed fowl similar to Yokohama.
Main Colours:
Black, White, Dark Reds.

Delaware

USA breed with Columbian markings.
Main Colour:
As for Light Sussex.

Dominique

USA breed with rose comb and full tail.
Main Colour:
Barred in a Slatey Blue and White.

Dutch Bantam (Old Dutch Bantam)

A tiny, cobby bantam, which carries its wings low, with a carriage similar to the Rosecomb bantam, although tail and other features not as well developed as the Rosecomb breed.

Main Colours:
Over 10 varieties in USA.
In UK and Holland Partridge, Silver Duckwing, Golden Duckwing and others are seen quite frequently and in Britain is a breed which is being taken up by many fanciers.
Guide to varieties (colours) see Old English Game.

Drawings courtesy Dutch Bantam Club.

FAVEROLLES*

This breed is rather similar to the Dorking in build and the fact it also has five toes, but with feathered legs.. It is bearded (muffling).

Main Colours:

Salmon: cherry mahogany shoulder, with a creamy hackle and tail hackle, with black breast. (See illustration).

Ermine: White with black hackle and tail ***(Columbian).***

Other colours, which are self-coloured, without any other colour are **Black** (beetle green), **Blue** (darker lacing), **Buff,** and **White.**

Legs should be white in the lighter colours, but dark in the Blacks and Blues. Eyes should be dark to hazel depending on body colour.

Faults: Smutty or faulty colours.

HAMBURGH* (Hamburg)

A Gamey looking breed of long standing, with rose comb with leader, and large curved tail.

Main Colours:

Black: Overall black, but not in bantams in UK, but exists in USA. Rosecomb in UK is the equivalent.

Pencilled - Gold and Silver:

Spangled - Gold and Silver

Special Notes: ***The markings are of vital importance; therefore must receive much attention.***

Colours: (Applies to Spangles & Pencilled)

Gold Ground Colour: a rich, bright golden bay or mahogany.

Silver Ground Colour: Silver white.

Markings: Gold: Tail, stripes, spangles and tips all a green/black.

Silver: Each feather with spangle of green black, (Spangles) or bars of green/black (Pencilled).

*** See Plate, page 48, for example of main colour or one of colours.**

Black   Silver Pencilled

Silver Spangled

Hamburghs

Note: The markings are quite precise so reference should be made to the illustrations of feathers.

Spangles: These must be round and on the end of each feather. Those which are V-shaped and run into the shaft of the feather should be discouraged. **Many modern-day birds have very large spangles running into each other which is incorrect.**

Pencilling: This is really fine ***barring*** which should be even. This is caused by an ***autosomal barring gene*** and is dominant. Tails should be laced. In the cock; the body is a Golden Bay or Orange-red (but **Silver colour in Silver Pencilled**) and, on the wing coverts, the upper web is black, and the lower web is pencilled with black. The underparts are also Gold (or Silver), slightly pencilled with black.

The overall impression of the Gold cock is of a beautiful, orange-red colour, with the wing bar edged with black, and slight black markings below the wing bay.

HOUDAN*

A five-toed breed, exotic in looks, with crest and beard, and a very unusual leaf comb.

Main Colour: Greeny-black with mottling in white.

INDIAN GAME*

One of the heavy weights of the fancy, low in carriage and very broad, with a pea comb and yellow legs.

Main Colours:

Dark: Male predominantly black with brown markings with chestnut wing bays. Hen is a deep chestnut colour with black double lacing.

Jubilee: Instead of black as for Dark, substitute White, possibly with some reddish tinging or chestnut.

Blue Laced: Blue with lacing as for Darks.

In the USA called Cornish, with many more colours.

*** See Plate, page 48, for example of main colour or one of colours.**

Jubilee Indian Game

LEGHORN*

The best known of the Mediterranean breeds of layers. Very active with upright comb in cock and a folded comb in hen, falling to one side.

Main Colours:

White, Black and Brown (in bantams), but other colours possible with wide range in Large: Pile, Duckwings, Buff, Blue, Cuckoos, etc. In fact, in the USA there are 23 varieties which include such exotic descriptions as Black-tailed Red and even a Millefleur, which is interesting but appears unnecessary for a utility fowl.

Cuckoo: See illustration: Light blue or grey with darker bars, not precise like Plymouth Rocks.

Buff: Sound, even buff, although some variation is found.

Brown: Black Red type so name is misleading. Male light red, but female dark partridge so double mating necessary.

Pile: Whitish birds, but with red hackle and shoulders in cock (See Modern Game Pile on Colour Plate)

Duckwing: Silver and Golden: Follows the traditional Duckwing pattern with Gold hackles and shoulders and remainder black for the cock and a grey colour with partridge markings for the hen. The Silver is lighter in colour.

DOUBLE MATING

For exhibition birds special breeding problems arise in Blacks and Browns. Whites can be bred fairly easily to standard type.

Double-mating **(one pen for pullets and another for cockerels) is necessary for Blacks, because the natural leg colour is black or grey; thus:**

(a) Because legs are yellow there are difficulties with dark under-colour so mating must allow for the fact that under-colour in males tends to be weak and stronger in females.

(b) Cocks with white in sickles tend to produce better pullets be-

---

*** See Plate, page 48, for example of main colour or one of colours.**

Some of the Colours: Leghorn

Cuckoo shown on Colour Plate: Black & White common colours.

cause there is a tendency for these birds to have strong yellow pigment.

(c) Pullets have stronger black, but the shanks do not come as yellow as the male; they tend to be a dark colour. Accordingly, the cock with white in the sickles (less dense black plumage) will tend to produce better yellow in the legs of the pullets because the overall density will be reduced.

Browns have problems with a ruddiness across the wings which come from a very bright cock. In the male the desire for a pale yellow hackle, yet with a black breast and tail, does cause conflict. Males and females must be selected very carefully to eliminate these problems.

Combs can cause difficulties - the fold for the hen and the straight, strong comb for the male. Breeding pens should allow matings which emphasize the separate requirements; eg, hen with erect comb to breed cockerels and weak comb on cock for pullet breeding.

NANKIN BANTAM*

Description: A well rounded bird with wings fairly low; prominent breast and a fairly upright stance; large sweeping tail gives symmetry. Active, with a proud carriage and beautiful feathering.

Main Colours:

One colour, a combination of ochre and cinnamon. Takes its name from the yellowy colour; ie, Nankin is a colour name.

Legs & Feet: Clean legged with four toes; fairly short. Colour slate blue or white with a bluish tinge.

Beak & Eyes: Colour white; originally the *standard* (Entwisle) stated short and small, but the modern *standard* now stipulates ***longish and fine.***

*** See Plate, page 48, for example of main colour or one of colours.**

Chanticler (See page 60)

Ixworth

Some Rare Breeds

Note the similarity, both being developed from Indian (Cornish) Game.

OTHER BREEDS - NOT SHOWN IN COLOUR

Frisian

Similar to Dutch bantams in stature for bantams, but large fowl also exist.

Main Colours:

Pencilled in Gold and Silver (similar to Hamburghs), and **Black**. Other colour(s) may exist in Germany.

Frizzles

See Bantams, Colour Plate *Description* later.

Ixworth

An Indian Game–type breed developed by Reginald Appleyard the famous duck breeder.

Main Colour:

Only found in White and now quite scarce in large and bantams, which were always rare.

Java

In type and carriage almost identical to Jersey Giants; the latter were developed from Javas.

Main Colours:

Black and **Mottled** (rather like Anconas).

Jersey Giants

A very large breed which evolved in the USA from Brahmas, Javas, Indian Game and Langshans.

Main Colours:

Black, and **White**. Dark Legs.

Kraienkoppe

Lakenvelder

Two Foreign Breeds

Kraienkoppe

Of Dutch or German origin which is found in large and bantams. They look like the old type Old English Game, which had large and full tails.

Main Colours:

Silver (similar to Grey in OEG) and **Golden** (is A variation of the Black Red).

Lakenvelder

A German breed of conventional shape with a fairly large tail.

Main Colour:

White body with **black** neck hackle and tail.
See Columbian ***Description*** for relationship.

Lamona

A USA fowl, rather like the Dorking.

Main Colours:

Buff, Black, Columbian, Dark Brown.

Langshan

See **Croad Langshan**. The USA type is similar to the British ***Modern*** Langshan which has long legs, rather like the large Modern Game.

Main Colours:

Black,
White.

Legbar

A barred breed which was developed for sex linkage (ie, automatic sexing of chicks by colour). Campines were involved.

Minorca

A racy Mediterranean type; upright comb male and folded comb female. White ear lobes are important in getting the correct type.

Main Colours:

Black,

White, and

Blue.

Other colours have existed; there is a Buff in bantams in the USA.

Legbar

One of the Sex -linked breeds, where sex can be detected by the colour of the chicks

MALINES*

Tall birds with slightly feathered legs and a single comb. At one time fairly popular they are now a rare breed. The old name was ***Coucou de Maline,*** indicating the Cuckoo colour. However, this table fowl from Belgium is believed to originate from Asia; it has probably been developed from one or more of the breeds from that part of the world.

Main Colours:

Cuckoo: This is the typical ***cuckoo*** with bars which are irregular. (See Coloured Plate illustration).

Ermine: Columbian type with white body and black in hackle, wings and tail. See Light Brahma and Light Sussex.

White, Black, and **Blue**, all self colours.

Gilded Black, Silvered Black, Gilded Cuckoo, all now non-existent.

Legs are lightly feathered and are white in colour.

There was also a "Turkey-headed variety", with the same colours, so named because they were thought to resemble a turkey in the head, with rose comb, large gullets, and little or no wattles. Whilst this seems fanciful, they did exist.

MODERN LANGSHAN*

This was a breed which evolved from the Croad Langshan. It is a very large breed in large fowl with long legs, greyish in colour, which have light feathering on the side.

Main Colours:

Black with the most brilliant of green sheens.

Blue and **White** are the remaining colours. These colours must be brilliant, and shiny.

*** See Plate, page 48, for example of main colour or one of colours.**

Comparison of Hamburghs & Related Breeds

MARSH DAISY*

This unusual breed was said to have been developed to thrive on marshy soil, hence the name. The comb is a rose type with a pointed leader.

Main Colours:

Wheaten or Gold: Male: Gold of different shades, with a breast that is brown or stone coloured. The female is a dark wheaten colour with a cream breast. (Note: different shades of colour did [and do] exist).

Black, Buff and **White**, self colours, although the Buff sometimes has black in the tail.

Brown: This appears to be rather like the Old English Pheasant Fowl, which the breed resembles in many ways.

Legs should be a ***pale willow*** and the eyes reddish with a darker pupil.

OLD ENGLISH PHEASANT FOWL*

This ancient British breed is useful, yet quite exotic.. In type it is a traditional (Jungle Fowl) shape with a large curved tail. The rose comb has a pronounced leader.

Main Colours:

Gold: Male an overall golden red colour, with black striped hackles and spangles, as well as black tail feathers which may be edged with gold. Female is along similar lines. The black should be green–black with a brilliant sheen.

Silver: Whitish colour with black markings.

Legs should be slate blue and the eyes should be a brilliant red. Ear lobes must be brilliant white.

Note: It is useful to compare the Spangled Hamburgh which is a close relative.

*** See Plate, page 48, for example of main colour or one of colours.**

ORLOFF*

Often referred to as the Russian Orloff this is a breed with doubtful ancestry. In German literature it appears to be regarded as a type of Game fowl, no doubt due to the resemblance to the Malay (one side of its parentage). The bearded ancestor is unknown and lost in antiquity.

Main Colours:

Spangled: Follows Black Red pattern with even spangling (see Colour Plate illustration).

Black: With a beetle green sheen.

White: Self white with lustre to the feathers.

Mahogany: In effect a largely red variety; the female to have peppering of black.

Legs deep yellow and eyes orange or red.

Orloff Hen

(Different type to the Colour illustration)

*** See Plate, page 48, for example of main colour or one of colours.**

ORPINGTON*

This breed holds a special place in poultry circles, being a brred developed by Wm. Cooke of Orpington and then extended into a variety of colours.

Colours require to be watched very carefully. **Blacks** should be jet black, but with a green sheen; avoid bronze, purple and any barring, spangles or serious colour or marking deviations. Accordingly, both sides of the breeding pen should have sound colouring, although red tinges in the hackle do not matter provided the extra green sheen is present.

Buffs should be a golden buff colour through to the skin. Theoretically some variation is allowed, but the birds which win are a deep, even colour. **Blues** are blue with a darker colour on hackle and, on the cock, the shoulders and tail.

Other varieties were in vogue in large fowl, but not in bantams. These include the **Jubilee** Orpington (introduced for Queen Victoria's Jubilee) and the **Spangled;** the former (Jubilee) is similar to the Speckled Sussex; ie, is a Spangle which is a ***dark*** Black-Red Partridge colour (see Old English Game), whereas the so called Spangled is a Black plumaged breed with white spangles - rather like an Ancona. This is an anomaly in itself, because this is not the normal description for a Spangle.

Whites should be pure white with no other colour.

NOTE: The Orpington has varied tremendously from the utility type developed by William Cooke. It is now a very feathery breed, especially in Blacks, with a very short back. In the same way the amourt of shank showing varies a great deal, depending on fullness of feathers. Some of the other colours follow the traditional pattern.

Jubilee Orpington
Very similar to the Specked Sussex (see Colour Plate)

Spangled Orpington
This is really a *Mottled* variety (not Spangled)

Orpington Colours

White Orpingtons

Buff Orpingtons

Other Orpington Varieties

Modern Game

A tall, finely limbed bird which was evolved from crossing Malays, Aseel, and Old English Game for large Moderns. The bantams came from a similar crossing, although this is not known with exactness. These are illustrated on the Colour Plates .

Main Colours:**

Black Reds (Partridge & Wheatens), Piles, Brown Reds, Golden Duckwing, Silver Duckwing, Birchen (Greys),
There are many other colours available in the USA in bantams.

Norfolk Grey

Resemble Old English Game, but Leghorn blood has given them a deeper body.

Main Colour:

Grey, consisting of black breast and tail, with silver grey hackle, shoulders and tail hackle on cock, and overall blackish colour with silver striped hackle on hens.

Note: Follows the pattern of Grey OEG, Silver Sussex and Birchen Modern Game. See OEG Brown Red (Colour Plate) and substitute silver grey for red)*.

** For a full description and analysis of Modern Game Colours See *Understanding Modern Game*, Bleazard & Batty, available from the publishers.

* See Plate, page 48, for example of main colour or one of colours.

Two British Breeds

MARANS*

A ***utility*** type breed which lays deep brown eggs. Developed from a number of breeds; named after the town of Marans in France.
Main Colours: All Cuckoo Colours: Dark, Golden Cuckoo, Silver Cuckoo. These are simply variations in terms of depth of colour. There is also a **Black** with beetle green plumage.
Legs White; **Eyes** red bay.
Faults: Yellow tinge to plumage. White in Black variety.
Note: This is a utility type bird renowned for its deep brown eggs. Recently perfectly ***barred*** birds seem to have been developed, which is fine provided the usefulness is not lost. The *standard* allows proper barring; that aside, it is still a heavy, dual purpose breed.

NAKED NECKS*

Name comes from the fact the breed has no neck feathers. It is a rare breed and in recent years has been revived.
Main Colours: Black, White, Cuckoo, Red, Buff, Blue, etc.
The bare neck is a reddish colour which appears scaly.
Legs: yellow or horn; **eyes:** orangy.

NEW HAMPSHIRE REDS*

A utility type breed with large and bantams.
Main Colour: There is one variety with the male reddish bay or chestnut with a black tail. The female is similar with the lower neck feathers having black tips.
Legs are deep yellow with a red tinge; eyes are red bay.
White in plumage is a serious fault.

*** See Plate, page 48, for example of main colour or one of colours.**

NORTH HOLLAND BLUE*

A Dutch heavy breed which lays tinted eggs, and a utility type. The standards recommend that the useful properties, rather than exact barring, should be given prominence. The legs are lightly feathered.

Main Colour:

Blue–Grey

Barred with alternate bars of blue-grey and greyish white. Strictly this is more of a ***Cuckoo*** than a Barred breed.

The cock is a lighter shade than the female.

Legs are white as well as the skin.

LA FLECHE*

A tall breed with a very unusual comb, consisting of two prongs or spikes. It has a full tail. THe carriage should be bold and upright.

Colour: Black with a green sheen.

Legs are black or leaden black.

Eyes: Black or red is standard.

Some regard this breed as being very similar to the Black Minorca, but with the horn–shaped comb. The head points should be a bright red free from any white and the ear lobes should be pure white and almond shaped.

*** See Plate, page 48, for example of main colour or one of colours.**

MALAYS*

This tall, very hard feathered breed has been kept for very many years, despite its lack of utility properties. The tail slopes downwards. The important point about Malays is that they should conform to type so ***exact colouring*** is secondary; and is given only 6 points in the British standard.

Main Colours: A wide varieties of colours, but without the clear division usually found in other breeds, possibly because these has not been 'perfected' to the same extent so the varieties tend to be mixed. The author's experience in breeding Malays suggests that a number of colours are likely from birds which appear to be the same variety. In other words, they are rarely stable for exact colour.

Black Reds: The hens may be Wheaten or Clay or even Grouse coloured.

Blacks: Self Black.

White: Pure White.

Spangles: Burgundy cock with black breast; hen dark partridge. The birds should have an even scattering of spangles or spots over the body.

Piles and other Game colours also exist and follow the OEG pattern, but not exactly.

Eye colour is important and must be light such as light yellow, white or pearl.

Legs are a bright or deep yellow.

*** See Plate, page 48, for example of main colour or one of colours.**

Black Columbian

Buff White

Plymouth Rock Varieties

PLYMOUTH ROCKS*

This American-originated breed is noted for its utility properties, including egg laying. It is unusual in carriage because the head position is way above the top of the tail, giving it a shape referred to as the 'gravy bowl' shape.

Main Colours:

Barred: a bird with black bars across on a blue tinged white background. The bars must be narrow, but quite positive and not merging with the white, into cuckoo. The slowness of maturity is said to be one of the factors in achieving the high standard of barring; however, this is not the only reason – many years were spent in improving the barring. In addition, double mating was practised to improve the quality.

The gene is dominant and is sex linked and acts in a negative fashion by restricting the black and thereby producing whitish bands between the black bars.

Black which has a beetle green sheen.

Buff which is an even golden buff throughout.

Columbian with white body and black markings on hackle wings and tail.

White: pure white all over.

Legs should be yellow and eyes a reddish bay.

POLANDS* or Polish

This breed is quite ancient and has figured on old, classical paintings, being very attractive with its crest and beard,++ horn-like comb, and unusual multi-colouring.

The Crest: this is a very distinctive feature and must be of a stated colour; the breed has a raised skull for the crest.

++ In fact, some have no beard or muffling and have visible wattles, so there are two distinct sub-varieties.

*** See Plate, page 48, for example of main colour or one of colours.**

White Crested

Variety of Polands

Main Colours:

Without Muffling: Black, Blue, Cuckoo.

Muffled or Bearded:

Laced: Chamois, Silver, Gold, with lacing.

Self colours: Black, Blue and White.

White Crested with Wattles:

Crest white with colour band at front same colour as body. Body colours: **Blacks** – Green black with plenty of sheen; **Blues** – slate blue, self colour or laced; **Cuckoo** – even barring slate or light grey.

The 'Blue' was regarded as being slate, self blue or laced (Andalusian-type), but there appears to be a movement towards a self colour (UK).

Bearded Polands (Colours which follow may appear in non-bearded – ***USA standards***):

(a) **Laced** and

(b) **Self Colours:** Black, Blue and White.

Chamois: Rich golden buff in main colour with lacing on each feather in white or light cream; crest same colour; beard buff laced with white; tail laced; Hen is buff with lacing all over.

Silver (laced): Silver-white colour with black lacing; crest black at roots; hackles and primaries tipped with black; tail laced with black. Female silver white with lacing on each feather of lustrous black.

Gold (laced): As for ***Silver***, but replace silver white with rich red bay.

* **See Plate, page 48, for example of main colour or one of colours.**

REDCAP*

This is an old breed which originated in Derbyshire and is related to other breeds like the Old English Pheasant Fowl, and Hamburghs. The main feature is the very large Rose comb with a leader; ie, a long point at the back of the main comb.

Main Colour:

The male is a Black Red with fringes and tips in black on the red feathers. The female has a half-moon spangle on each feather. The black should beetle green. Some variation is found in shade and there is a tendency to prefer the darker coloured plumage. (See Colour Illustration).

Legs a grey colour and the eyes red.

RHODE ISLAND RED*

Another breed which originated in the USA , which may be single (the usual) *or* rose combed.

Main Colour:

This is really a black tailed Red. The main colour is a chocolate red, quite even, with a brilliant overall gloss.

This is an unusual breed for colour. Apparently the cock gives the necessary black and the hen the deep red++ and no good specimens can be bred without taking this fact into account. Double mating is not usually advocated, but the possibility that it is practised should not be overlooked.

Legs should be deep yellow and eyes red.

Rhode Island White: a white variety has existed, but was never popular.

++See *Bantam Breeding & Genetics*, Fred Jeffrey, where the author quotes from the experiences of a breeder.

* See Plate, page 48, for example of main colour or one of colours.

SCOTS GREY*

This is a native of Scotland, where it still thrives in large and bantams. The shape and size indicates some connection with Old English Game.

Colour: Cuckoo. The basic ground colour is blue white and light grey. The bands should be black and straight across on body, thighs and wings, whereas they tend to be angled on other parts. Sometimes it is suggested that Barred Plymouth Rock and Scots Greys are the same in barring. This is not strictly true. In Plymouth Rocks the very bold black barring should be relatively straight and even, but the Scots Grey is more crescent-shaped (curving), and the colours are lighter; the spaces are a light blue with bars less definite.

Scots Grey Bantams

SUSSEX*

The Sussex breed is one of the oidest breeds and is closely related to the Dorking.

Main Colours:

Light Sussex: Columbian-type plumage, which is the most popular variety. The overall colour is white with the hackles striped greenish-black; the tail is also black, with the lesser coverts laced. Wing feathers also have some black markings. The neck hackle black stripes nowadays tend to be very broad and black, yet it must be appreciated that there should be a white ***margin*** on the edge of each feather. Eyes orange.

Brown: A rich mahogany colour with the black striped hackle and black tail. The female is a browny Partridge colour, with a pale breast of wheaten brown. It is really a Black-Red type. Eyes red.

Buff: In markings like the Light Sussex, but the white is replaced with a good, deep, even buff. Eyes red.

Red: As for Lights, but a deep red as the main colour and, once again, striped hackles. The under-colour is slate. Because of the colour this variety is sometimes said to resemble Rhode Island Reds, plus the striped hackle; whilst this is true the shape should be different, the Sussex being deeper with a very broad back. Also legs white in Sussex.
Eyes orange.

Whites: These are white all over. They appear to be out of place with the other varieties because they do not have the striped hackle. In fact, they are not very popular.
Eyes orange.

Silver Sussex Bantam

Buff Sussex

Sussex Colours

Silver: These are similar to the Grey found in Old English Game being blackish, but with neck hackle, shoulders and tail hackle a silvery grey (male). The female is black with a silver striped hackle and a laced breast, the breast being a little lighter than the rest of the body. Eyes are orange.

Speckled: There is a suspicion that Spangled Old English Game, are in the original makeup. They are of course a form of Spangle and follow the colour pattern for that variety, except the body colour is a dark burgundy colour for male and female. The ground colour is dark mahogany with a brilliant sheen on the male. The speckles should be spread in even fashion and there should be a small black bar separating the white speckle from the normal colour. With age there is a tendency for too many speckles to appear as well as too much white in the tail of the cock. Eyes red. See Coloured Illustration.

Leg colour should be white

Light Sussex Bantams

*** See Plate, page 48, for example of main colour or one of colours.**

Brown Sussex

Sussex Colours

Red Sussex

Light Sussex

More Sussex

Rheinlander

Shaped like the Hamburgh with a rose comb along the lines of the Wyandotte. This is a German breed.

Main Colours:

Black, White, Blue, Barred, Partridge.

Scots Dumpy

Very short legs and long body; belong to the Creeper category mentioned under German, Japanese and other short legged breeds.

Main Colours:

Dark Grey, Silver Grey, Cuckoo, and Black.

Shamo

A Japanese breed, although it may have originated originally in China. In type a mixture of Malay and Aseel.

Main Colours:

Black Red, Dark Red, Duckwing (Gold & Silver), Buff, Brown, Spangled, Black, Red, White and various off colours.

Sicilian Buttercup & Sicilian Flowerbird

Although not usually regarded as Mediterranean breeds they do have similar body lines to the Leghorns, Anconas and Minorcas - possibly not as streamlined. The extraordinary feature is the cup comb with points around the outside, rather resembling a flower. Size around 6lb for Buttercup and 4lb for Flower bird. Bantams 25 per cent these sizes.

Main Colours:

Buttercup: White, Brown (Black Red type), Golden, Silver, Duckwing; **Flowerbird:** Mahogany, and Spangled.

Sicilian Buttercup

Rheinlander

Two Unusual Breeds

Shamo Large Fowl

Bantams do not follow the upright stance of the large fowl

SILKIE*

A small breed which is found in large and bantams and which has some unusual features. It has a dark face and turquoise ear lobes; some have crests and, additionally, the others beards as well as the other features..

Main Colours:

White, Black, Blue, Partridge and Gold.

Colours which are *self* should be the colour stated without any other mixture.

Whites are pure white, without any other colour mixed in, or yellowy in tinge.

Blacks are greeny-black. It is suggested that 'colour in the hackle' is permissible (British standards), but this cannot be read too literally or the variety is no longer a Black.

Blues are an even colour without lacing; a darker top colour is permissible, but not too dark so as to become black.

Golds are a deep golden buff, evenly distributed.

Partridge are rather like badly defined Black-Reds with the cock being golden-red with black breast, and the hen a golden-brown with detailed mixtures of black and brown resembling a Partridge (see Black Reds under OEG). This colour requires a great deal of work to be done on it to become a true Partridge or we must accept that the variety will never be true Partridge in the sense of having a distinct pattern to the female feathers. The nature of the feathers, ***down*** instead of proper webs, does not allow intricate patterns to be developed.

Legs should be a lead colour (dark grey) and the beak to match (slate). Eyes should be black and the face a dark colour known as mulberry, with the ear lobes a turquoise blue colour. It is a serious fault to depart from these colours.

*** See Plate, page 48, for example of main colour or one of colours.**

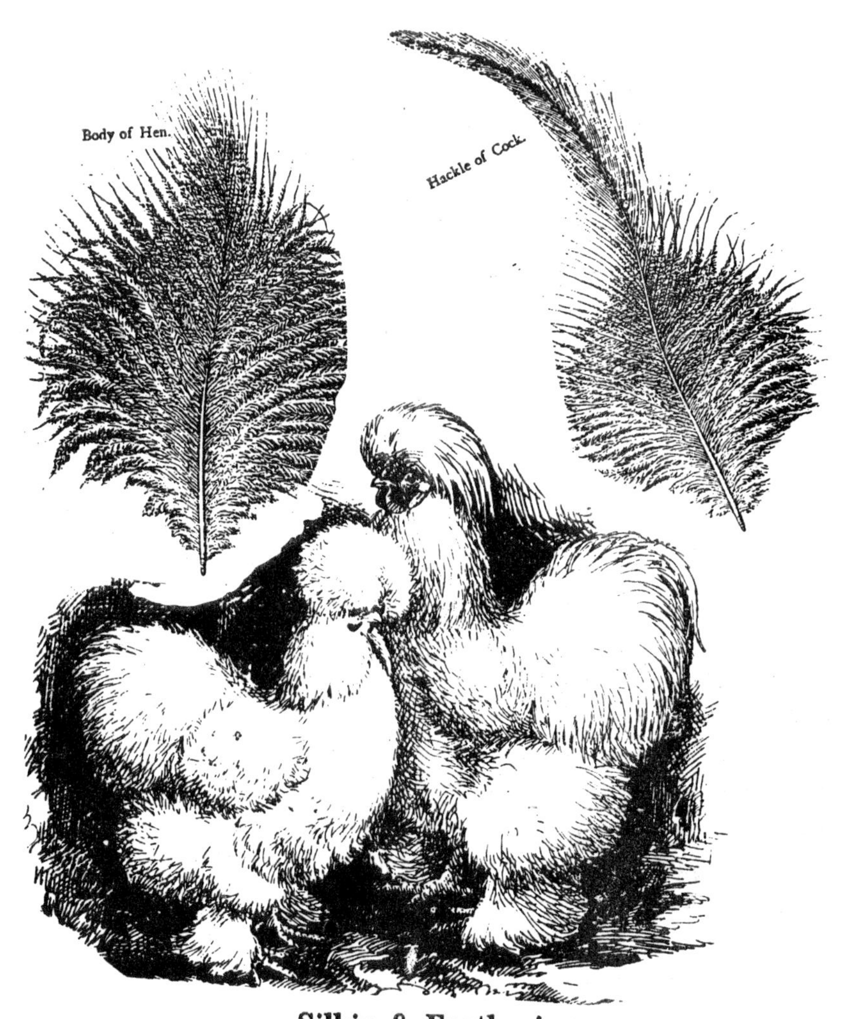

Silkie & Feathering

This webless feathering make exact patterns difficult to create in different colours.

Note: For more detail on this breed see *The Silkie Fowl,* J Batty, available from the publishers.

WELSUMMER*

A utility type breed, developed for its dark brown eggs which is, nevertheless, very attractive. It has a moderately upright carriage and well feathered.

Main Colours:

"Partridge/Black Red"

The cock resembles a conventional Black Red, somewhat "mismarked", because the breast has brown mottles. The hen is a rich Partridge colour, but with a breast which is deep red (instead of normal salmon), and her tail is black with brown pencilled outer feathers. This deepness of colour is no doubt the reason for the brown in the breast of the male, and the rich golden brown of his hackle.

Silver Duckwing which follows the conventional pattern with silver hackle and shoulders, with the mallard-type bar across the wings (See Old English Game). Some mottling is allowed on the breast.

In both cases the breast and tail and other black areas should be greeny black.

Legs should be yellow and eyes red. The beak should match the legs.

*** See Plate, page 48, for example of main colour or one of colours.**

YOKOHAMA & PHOENIX*

These are of Japanese origin and in their own country are known as ***Onagadori*** in large fowl, with exceptionally long tails, up to 30ft. long. In bantams the tails are still long when compared to other breeds, but are at the most a metre, and usually much less. In Britain and other countries the breed has a much shorter tail and known as Phoenix or Yokohama.

They are said to be pheasant-like and certainly, with their long bodies and large tails the overall impression is that of a Reeves or other of the large ornamental pheasants.

Combs are single or walnut or rose and at the time of writing the British ***standards*** give all the varieties under the group name of ***Yokohama***. However, in the older standards the single combed birds were the Phoenix and the Yokohama had walnut combs.++

Main Colours:

Walnut varieties are either Red Saddled or White.

Single combed birds may be any colour found in Game and therefore are seen in Duckwing (Silver and Gold), Spangled, White, Black Reds and others.

Bantams are around, but quite scarce. They have been shown from early times, but never seem to become popular. This is a great pity because the Yokohama is an extremely attractive breed. Weights in bantams should be in the region of 22 to 26oz. (APA).

*** See Plate, page 48, for example of main colour or one of colours.**

Illustration of Japanese Onagadori

WYANDOTTE*

Originally developed as a utility type fowl in the USA, but the bantams came from different crosses. Has an unusual rose comb, somtimes referred to as a 'cradle comb'. presumably because it looks like a cradle upside down.

Main Colours: Barred, Black, Blue, Blue Laced, Buff, Buff Laced, Columbian, Gold Laced, Partridge, Red, Silver Laced, Silver Pencilled, White. **In the USA there are 18 varieties in bantams:** Barred, Birchen, Black, Black breasted Red, Blue, Blue Red, Brown Red, Buff, Buff Columbian, Golden Laced, Lemon Blue, Partridge, Silver Laced, Silver Pencilled, Splash, White, and White Laced Red.

Past Colours: Buff Columbian, Black and White Spangled (Ancona type), Pile, Black-laced White, Copper (Golden Duckwing type), Cuckoo (similar to Cochin), and Mottled. Some of these have been revived.

COLOURS:

Laced:

Silver: Very handsome; Silvery white with black stripe in neck hackle; each feather laced in a regular fashion.

Gold: Golden brown ground colour; and as for Silver.

Blue: Red brown with blue lacing. In male hackles and saddle are darker.

Buff: Even, rich buff colour with white lacing. Hackles striped with white down centre. White under colour.

* **See Plate, page 48, for example of main colour or one of colours.**

LACED

Laced means there is an edging around the feather in a different colour. In Wyandottes it is single, but other breeds have double lacing; eg, Indian Game hen.
Faults are: 1. Horse shoe lacing (partial only). 2. Full or part double lacing. 3. Mossy or streaky feathers. 4. V-shaped lacing (instead of rounded). 5. Irregular line of black.

Pencilled Wyandottes

Pencilled in this connection means fine crescentic (curved) marks on each feather, in the form of lacing, often in triple form and varying in width and darkness of colour, depending on the type of feather.
Partridge: Male is a Black red type with head of red and the rest of the hackle becoming lighter through orange to lemon at the base, with a black stripe on each feather. Top colours are red as in a Black Red. The breast, legs and tail should be green-black.
Female: Ground colour is Partridge, the colour of a dead oak leaf, with concentric rings of black on each feather.
Silver Pencilled: as for Partridge, but a silver-grey ground colour.

Differences in British & USA Standards

The USA standard (APA) treats many breeds under the heading of 'Partridge' as if identical, but Wyandottes are *not* the same as Plymouth Rock Partridge in the British standards. The female Wyandotte should be the colour of a dead oak leaf; ie, Partridge resembling OEG (although not exactly the same), but in APA the ground colour is ***deep reddish bay*** and the illustration in the Book of Standards shows this dark red colour not partridge. The British standard for Partridge Plymouth Rocks follows the USA standard (APA).

Example of Silver Laced Wyandotte Pullet

Note: this is a young bird not fully matured.

Cockerel breeding pair, ie; male no lacing on shoulders;exhibition type. Female large with sound black lacing & silvery hackle.

Pullet Breeders. Male sound under colour with distinct lacing; hen dark neck hackle & pure yellow legs.

Laced Wyandottes (Double Mating)

OLD ENGLISH GAME*

Type: Agile, aggressive with other birds, but with distinctive characteristics; rather like Red Jungle Fowl modified by selected breeding. Different types exist, namely, the so-called Oxford and Carlisle types, but this is an artificial division. Bantams are a different type altogether.

Origin: England, but kept in Scotland and Wales for a very long period. Some doubt that the bantam race are miniatures of large Old English Game, but there is no proof either way, except they do resemble the large Game in many respects.

Size: Size in practice varies from around 2 to 3 kilos, but should not be too large. Bantams are small and cobby: female 18 to 22oz. (510 to 620g.) and males 22 to 26oz. (620 to 740g.).

Hard or Soft Feathered: Hard feathered, feathers fitting close to the body.

Comb: Single and in males usually trimmed (dubbed); there are Modified versions include birds with Muffs and Tassels.

Tail: A full tail in large OEG, but in bantams the English-type birds have very small tails with little or no main sickle feathers. In the USA and Australia they prefer the older type which have large, fully sickled tails, like large OEG.

Legs & Feet: Rounded legs, finely scaled, colour to match plumage; four toes with back toe firmly on the ground and **no hint of being duckfooted**. Legs in bantams usually white, except in darks such as Greys and Brown Reds. In large a wide variety of leg colours to match pigment content of variety. Partridge mamy have white, yellow, green-willow, carp, grey and blackish.

Beak & Eyes: Generally white or horn to match legs. **Eyes:** Red, or dark for dark coloured birds.

*** See Plate, page 48, for example of main colour or one of colours.**

Colours: There are numerous colours and although it is often stated that in Game there is no such thing as a "Bad Colour", in reality, when a definite colour is specified, the bird in question should comply with the description in the standard. The more popular colours are:

1. Black Red with Partridge hens. This means cock with scarlet hackles and shoulders and tail and breast black. The hen is a soft brown colour with black and lighter markings which resemble a Partridge and the wings are free from rust colour. Some carry a green sheen and have a gold shaft on each feather.

2. Black Red with Wheaten hens. Cock as 1. but a lighter, more orangy shade for the scarlet. The Wheaten is a light creamy colour the colour of wheat.

3. Spangle. The male a deep burgundy colour with black breast and tail and small speckles in white, evenly distributed over the body. The female is similarly marked.and is of a deep partridge colour with a small black line near each spangle.

4. Duckwings. These present an anomaly because the description comes from the presence of a mallard-type wing bar of metallic blue found on the wild duck. There are two main types:

(a) Golden Duckwing and, (b) Silver Duckwing.

Black breasted and yellow replacing red in Black Reds for Golden Duckwing and silver white for Silver Duckwing. Females rather like partridge hens, except the body colour is a darkish grey mixture for Golden Duckwing and light grey for Silver.

5. Self Colours: Black, Blues, Whites should all be pure in colour without foul feathering.

Face red or in Dark Reds a Gypsy (Dark, dusky colour)

Eye colour: Red preferred (very light eye unacceptable) Dark Reds black eyes.

Hackle varies; Light Reds (Wheaten Hens) have orange-red, but other shades exist. Dark legged birds have deeper colour.

Tail black with green sheen.

Breast: Black with no other colour. This differs from Welsummers which have 'marbling' in brown on breast.

Legs may be white, green, yellow, carp, grey or black to match colour variety.

Black Red Colours

6. Pile: Males are white instead of black in Black Reds, but with bright red on hackles and shoulders; the hens are white with a yellow hackle, spotted or shaded breast (creamy) and depending on the intensity of the red the overall colour is white or white with red or yellow markings. They are given names like Custard Piles or Blood-wing Piles based on the depth of colour. Legs should be white, but many are seen with deep yellow legs which *may* be a sign of an Indian Game cross at some point.

7. Furnesses and Polecats: A mixture of black, brown, red and yellow. In the Furness cock the shoulders are a red colour and a variation, Brassy-backed, the shoulders are yellow. Usually the hackle is black. The Polecat is a similar mixture. In all cases the hens are basically black with intermingling of another colour. Although the purity of colour is doubtful these are attractive birds.

8. Brown Reds and Greys: Basically black birds, although may have signs of dark partridge in the female feathers. Brown red cocks have a black breast (sometimes laced) with hackles and shoulders an orangy yellow and females have a yellow stripe in the hackle.

Greys are similar but substitute a silver grey for the orange yellow. Both are scarce in bantams. Legs and eyes should be dark.

9. Blue Variations: Once blue is introduced into any variety it has a tendency to stay. The result is that many sub-varieties have been produced, such as Blue Tailed Wheatens, Blue Reds, Blue Duckwings, Lemon Blues and many others.

10. Crele and Cuckoo: These are barred varieties, sometimes blue/black, like other barred breeds, and others red and or yellow, making a very beatifully marked bird.
11. Rare Colours and Off-Colours: Many other variations exist ranging from Ginger Reds and Black Breasted Dark Reds which are seen in large Game to off-colours that have been produced by a chance mating, when the main aim has been to produce better shape. These are also Black and White Splashes, Brown Breasted Reds, and many mixtures that defy accurate description. They have to be shown in ***Any Other Variety*** classes.

Note: For coloured illustrations of the many of the different colours see Understanding Old English Game, J Batty, available from the publishers.

*** See Plate, page 48, for example of main colour or one of colours. This is a Brown Red, being a darker version of the Black Red cock, but the hen is black, possibly with dark brown markings, with an orange or lemon hackle. Sometimes both male and female have brown lacing on the breast; this no doubt is a throw back to the original colour when the cock had a completely brown breast. It was from this fact that the variety got its name - brown breast and mainly red remainder, the fashion being to specify the breast colour in the description.**

Spangle Cock

Silver Duckwing Cock (Silver white & Black)

Other Large Game Varieties

Shamo

The Shamo is a large Game fowl which is said to originate from Japan, but may have have come from China many generations ago. In type there is resemblance to the Malay, but tend to be more upright and rather thickset.

Bantams of the name are known, but are not true miniatures of the large fowl, being more cobby and shorter than the large.

Main Colours

Any Game colour, but those seen are Black Red, Duckwings, Black, Pile and various off colours.

For illustrations of ***type*** see page 107 where there is shown a White hen, Clay hen and Black cock.

Shamo Bantams

Spanish

A Mediterranean type fowl with an extraordinary face adornment in the form of a long white attachmant stretching from the base of the comb to the wattles or below. See page 55.

Main Colours:

Black, but White and Blue have existed and may still be kept.

Sultan

A breed rather similar to the Polands, but with vulture hocks, and feathered legs.

Main Colours:

White, but Blue and Black have been bred. Off-colours are also seen.

Sumatra Game

Related to the long-tailed fowl of Japan this is a beautiful breed. It has a dark gypsy face and a rose comb. Although a Game breed and therefore hard feathered, in recent years the Rare Poultry Society have classified it as a long-tailed soft feathered breed. This seems illogical, but possibly it is regarded as being more comparable with long tailed breeds for exhibition purposes.

Main Colours:

Black with a brilliant, green sheen.
Whites and Blues have been known.

Thai Game

These are rather like the Shamo; large strong birds, rather excitable. Those kept by the author had hens which were almost identical to the Malay hens, but the cocks had more fan-like tails, with strong feathers.

Main Colours Similar to Malays; tend to throw a range of different colours - Black Red, Black and mixtures of these.

Thuringer Bearded

A very attractive breed from Germany and appears to exist in large fowl and bantams. The full beard makes them quite unusual. Bantams are available in the USA, but the breed does not seem to be bred in the UK, although the author remembers visiting Rex Wood at his rare breed centre many years ago when he had German breeds, including possibly the Thuringer.

Tuzo Bantam

A small, hard feathered bantam from Japan, similar to the Shamo bantam or Aseel.

Main Colours:

Any colour which appears in conventional Game varieties; eg, Black Red, White, Spangles, Black, etc.

Thai Game Cock

White Sultan

Sumatra Game

Green-black lustrous plumage, with gypsy face

Two Rare Breeds

Thuringer Bearded
Silver Spangled

Tuzo Bantams

Vorwerk

A German breed which is rather like the Lakenvelder in being one colour in the body and a different colour in the hackle and tail.

Main Colour:

Buff and Black combined. Rather like the Lakenvelder with black hackle, but buff body (instead of White).

The male should have a black hackle; the saddle colour should be buff with faint stripes, the rest of the body is also a buffish colour inclining towards a reddish colour. Legs are slate.

In the female the overall colour is buff, but not the even buff found in many other breeds, such as Plymouth Rocks of that colour. In fact, the breast and lower parts of the body may be a lighter shade, somewhat patchy.

Eyes are orangy-red.

The hens may vary in body colour, especially when subjected to severe sun and bad weather conditions.

Vorwerk Male & Female

BANTAMS*

As noted earlier, bantams may be counterparts of the large fowl (25 per cent of standard fowl) or may be ***natural bantams,*** ie, being a natural size as bantams, with no equivalent in a large size. These are also known as 'True Bantams', thus indicating that there is only the bantam size.

There are problems in using either term because some breeds do not readily lend themselves to either category; for instance, the Sebright which was a 'man made' bantam - is this a true bantam? Old English Game bantams are now quite distinct from the large Game and arguably are true bantams. Pekins are now a separate breed and regarded as natural bantams, but in the USA are still called Cochins like the large counterpart. Modern Game bantams appear to have been produced from a different source to Large Moderns, and it could be argued that they are true bantams.

Obviously then there is need for great caution to be exercised by not being too dogmatic in placing birds into definite categories.

As far as possible the bantams which have large breeds should comply as near as possible with those breeds, thus recognizing that they are supposed to be reduced versions. However, it is a fact that in many instances the bantams have reached greater perfection in the colouring achieved simply because bantams are now bred and shown more frequently.

***Footnote.**

Note: For details of the origin of bantams and a detailled description of all the main varieties the reader is referred to *Bantams & Small Poultry,* J Batty, available from the publisher.

RUMPLESS BANTAMS*

Although 'Rumpless' may be found in soft feather breeds as well as hard, the latter is more usual. In the standardized type they are rather like slightly oversized Old English Game, but with no tail. This is due to the end of the spinal cord being missing.

Origin: Uncertain, but known in Persia and various other countries, such as Japan and South America, hundreds of years ago. Aldrovandus described them in 1645, in the days of Oliver Cromwell. At one time it is said that they were common on the Isle of Man; another report stated Belgium and, in another, the West Indies.

Size: Small being 18oz for pullets and 26 for cocks (510 and 740g.).

Hard or Soft Feathered: Hard, although Rumpless are also found in Soft feathered birds. Araucanas can also be Rumpless (see entry).

Tail: None..

Legs & Feet: As for OEG; clean legged with back toe pointing backwards, and no hint of duckfootedness.

Beak & Eyes: Slightly curved and colour to match legs. **Eyes:** Red or dark, depending on colour of plumage.

Main Colours: As for OEG bantams, but not so strict.

The coloured illustration is the Golden Duckwing where the cock has a light yellow hackle, black breast, gold across the shoulders and a greeny-black or steel-blue wing bar from which the name Duckwing arises, being so like the bar on the Mallard duck. This is an important colour for OEG, both large and bantam; for the normal bird the tail is green-black.

*** See Plate, pages 48 - 49 for examples of main colour or one of colours.**

Variety of Colours in Rumpless.

PEKIN BANTAMS*

Type: A small, bantam with heavily feathered legs and very ornamental. Originated in China; at one point the Pekin (Cochin in USA) was regarded as a bantamized large Cochin, but opinion has now concluded that it is a natural bantam. Certainly on the immediate evidence, the robbing of a Palace in Peking, and the acquisition of the bantams by soldiers, appears conclusive, but China has a long history of poultry keeping and way back in history the birds may have been bred down. There may also be a link with Japanese bantams because these are every similar, although the ***Creeper*** characteristic (see Japanese) is not mentioned in Pekins.

Main Colours:

Black, Blue, Buff, Cuckoo, Mottled, Barred, Columbian, Lavender, Partridge, and White. **In the USA** there are 16 varieties (see ***Cochins***)

Black: Beetle-green black throughout.

Blue: Pigeon blue with darker hackles.

Buff: Even buff colour throughout.

Cuckoo: Dark slate barring on a french grey background colour.

Mottled: Black with white mottles evenly marked.

Barred: Definite barring of black/white, the colours being quite distinct.

Columbian: As for Light Sussex.

Lavender: A silver-tinted lavender (not a light blue).

Partridge: Male Black Red and female a light partridge colour (see OEG of same colour).

White: An ultra-white overall.

*** See Plate, pages 48 - 49 for examples of main colour or one of colours. For descriptions and illustrations of all colours see *Pekin Bantams*, Margaret Gregson, BPH.**

Comb: Upright and single, with even serrations. Face bright red. Ear lobes and wattles long and smooth.
Legs & Feet: Very short with the thighs covered in fluff and the shanks covered in feathers, as well as the middle and outer toes. Colour yellow.
Beak & Eyes: Slightly curved and small, yellow in colour, set into the quite small head. Eyes red, yellow or orange. Beak yellowish.

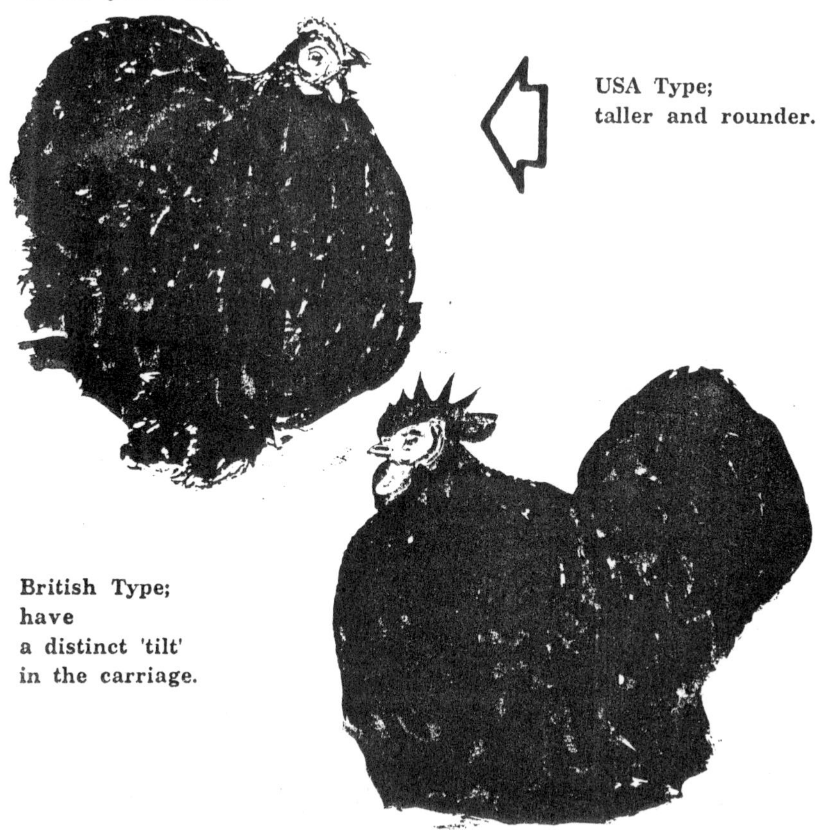

Silhouette of Pekin Bantams

FRIZZLE*

Type: Rotund bodied fowl, large and bantams, which have curled feathers in place of the conventional, straight feathers. However, they appear to be a distinct species as bantams. Early illustrations show bantams in their own right.

Origin: Asia possibly Japan or China.

Size: British standards 20 to 28 oz. (570 to 790 g): USA allows for cocks to go up to 30 oz.

Hard or Soft Feathered: Soft; feathers curl the wrong way.

Comb: Single.

Tail: British standard stipulates that tail should be large, full and erect, but in practice this does not seem to be the case. Modern birds have moderate tails much lower than the line of the head

Legs & Feet: Medium with very little, if any, of thigh showing. Clean legged in Britain, but in the USA there are feathered legged types, rather like the Pekin.

Eyes & Beak: Eyes full and beak fairly short.

Description: A small bird with distinctive feather growth, each one to be curled towards the head, in an even fashion.

Colour: Many colours, including **Columbian, Selfs** (black, blue, buff, white) and many of the **OEG colours** (Black-Red, Pile, Spangle, etc.) and **Rhode-Island-Red-type** Dark Reds. Many beautiful 'off' colours' have been bred and these are quite extraordinary.

* See Plate, pages 48 - 49 for examples of main colour or one of colours.

Selection of Frizzle Colours

JAPANESE BANTAMS*

Type: Short-legged dwarf fowl (coming into the category of 'Creepers' which have certain characteristics, discussed below), with a low carriage and upright squirrel tail. Wings touch ground. Natural bantams. They are to be found in normal feathering as well as ***Frizzles*** and ***Silkies.***

Origin: Japan where they are known as ***Chabos***.

Size: Small: 14/18 oz to 18/22 oz (400/510 to 510/620 g.) for male and female respectively.

Legs & Feet: Legs very short, **and thick**, with no feathers; feet barely visible; bird walks with faultering gait or waddle. They are yellow in colour.

Beak: Strong and curved; yellow in colour. **Eyes:** Orange or red.

Main Colours: These are numerous and there are many variations depending on the club concerned, eg; USA standards or British. Some are self explanatory:

> **Black Tailed White; Black tailed Buff; Buff Columbian;** (see description of Columbian Wyandotte, but replacing buff for white); **White; Black; Birchen Grey; Silver Grey; Dark Grey; Miller's Grey; Black Mottled; Blue Mottled; Red Mottled; Self Blue; Lavender Blue; Cuckoo. Red; Tri-Coloured; Black Red; Brown red; Blue Red; Duckwing and some 'off-colours'.**

*** See Plate, pages 48 - 49 for examples of main colour or one of colours.**

Characteristics: A cobby bird with a variety of features which make up its basic characteristics; short back; large tail; short legs; tiny body, and drooping wings.
Each bird carries the short leg and long leg genes together and, as noted, if the two shorts come together, the result is death. In effect therefore the **norm** will be one quarter lethal, the same for long leggedness, and half will contain both long and short leg genes.

Exhibition Faults: Any uncharacteristic feature such as short tail, long legs, long back, faulty comb, wings set high, **lobes** which are white, any other fault not consistent with a dwarf bantam.
Special Notes: Japanese are for dedicated show people; they present a serious challenge and there are many colours. Because they are what is known as a "creeper" type there are many breeding problems. Even if fertility is satisfactory the ***short leg X short leg*** is a lethal gene and this results in chicks dying in the shell, usually around the 18th day. Yet the correct procedure for breeding is to mate short leg with short leg, because ***long leg X long leg*** will only produce long legged birds.

*** See Plate, pages 48 - 49 for examples of main colour or one of colours.**

Japanese Bantams

OLD ENGLISH GAME BANTAMS*

Origin: England.

Size: Small, cobby: female 18 to 22oz. (510 to 620g.) and males 22 to 26oz. (620 to 740g.).

Hard or Soft Feathered: Hard.

Comb: Single and in males usually trimmed (dubbed); there are also varieties of ***Muffs*** and ***Tassels.***

Tail: In English-type birds very small tail with little or no main sickle feathers. In the USA and Australia they prefer the older type which have large, fully sickled tails, like large OEG.

Legs & Feet: Rounded legs, finely scaled, colour to match plumage; four toes with back toe firmly on the ground and **no hint of being duckfooted**. Legs usually white, except in darks such as Greys and Brown Reds. The large Game have a much wider variety of leg colours and the requirements are not as strict as for bantams where the wrong colour legs would result in a bird being 'passed'. Yellow legs are a bone of contention; for example, Piles are shown with white or yellow legs, but many judges prefer the former, which is regarded as being more correct - there is always a suspicion of a cross with, say, Indian Game, when yellow legs are present.

Beak & Eyes: Generally white or horn to match legs. **Eyes:** Red, or dark for dark coloured birds.

Colours: The more popular colours are:

1. **Black Red with Partridge hens.**
2. **Black Red with Wheaten hens.**

*** See Plate, pages 48 - 49 for examples of main colour or one of colours.**

3. Spangle.

4. Duckwings.

(a) Golden Duckwing and, (b) Silver Duckwing.

5. Self Colours: Black, Blues, and Whites.

6. Pile

7. Furnesses and Polecats

8. Brown Reds and Greys

9. Blue Variations

10. Crele and Cuckoo

11. Rare Colours and Off-Colours

Colour Plate (pages 48 to 49) shows a Black Red Partridge pair. The hen colour is on the dark side, a more even partridge colour being preferred. For more colours see *Old English Game Bantams* or *Understanding Old English Game,* J Batty.

The markings are rather like a mackerel (Blueish) or red, orange or lemon. Rather similar to Cuckoo markings, but more intermingled.

Crele OEG Bantam

This cockerel is rather long in the back.

Spangle OEG Bantams

Black OEG. The cock (right) has a very short tail, but good front; colour excellent - deep black.

OEG Bantams

SEBRIGHT*

Type: An ornamental type bantam which was developed from other breeds about 200 years ago. It has hen-feathering, delicate lacing and a rose comb.

Origin: The originator was an Englishman, Sir John Saunders Sebright, Baronet, Member of Parliament for Hertford and a keen animal breeder and falconer. He crossed the Nankin with Polands, a Hamburgh (and /or Rosecomb), and a Henny Game cock.

Size: Very small; 18 to 22 oz. (510 to 22g.), the cock being larger. USA standard (ABA) slightly heavier.

Hard or Soft Feathered: Soft, but feathers close fitting around the body.

Comb: Rose which is square at the front and tapers to a point at the back (the leader), turning slightly upwards. The comb is full of small serrations and should be even all over, without bumps or hollows. The face, comb, ear lobes and wattles should be red in the cock and mulberry or gypsy in the hen. In the original breed both sexes had very dark faces and, if possible, this feature should be the aim.

Tail: Hen feathered, which means male and female have similar tails so the cock has no sickle feathers. The tail should be spread out rather like one side of a fan and the feathers should be laced around the edges; poor specimens have no lacing or have spangles at the end of the feathers, or suffer from frosting or mismarking. Tail carried high, in the region of 70° angle above horizontal (ABA Standards).

Legs & Feet: Legs are fairly short with thighs wide apart. In colour they should be slate blue; four toes.

*** See Plate, pages 48 - 49 for examples of main colour or one of colours.**

Beak & Eyes: The beak should be dark horn, but, in the Golds, dark blue is allowed. Eyes quite full with dark ring round (cere), and a dark colour – brown or black.

Description: A jaunty breed with an upright carriage, wings pointed downwards at an angle, broad prominent breast, short back, short hackle, and tail with no sickles on male.

Colours: **Gold** and **Silver**. Other colours have been bred and in the USA some extraordinary varieties have been attempted, but nowadays the original colours remain – Gold and Silver. In fact, there seems little point in developing other colours.

Characteristics: An ornamental bantam much admired; jaunty and colourful, with many unusual features.

Show Qualities: Excellent, but difficult to breed birds with all the requirements to a high standard.

Exhibition Faults: Faulty lacing; uneven colour in Gold and creamy colour in Silvers; single comb or badly shaped comb; feathers not rounded or almond shaped; curved sickles on male; tail at a distinctly wrong angle; long back on male (some tolerance is allowed in female); white or wrinkled ear lobes; shaftiness in feathers; lack of character – not strutting and trembling when excited (especially male); long legs; any other trait or feature not in accordance with the standard.

Double Mating: The practice of having one pen for females and another for males is not essential, although some breeders do use special matings to get specific results. Since the aim is to get perfect lacing it is usual to mate on the basis of getting delicate, yet bold lacing. If trying to improve males it is usual to use a heavily marked male and a normal female. For female breeding the male should be well laced and standard type, with a female that is very heavily laced in all parts. Opinions differ on the exact approach, but many fanciers suggest that Silver and Gold should not be bred together, because this spoils the colours; in fact, at one time pure Silvers were almost lost because of mixed breeding. The **Gold X Silver** does provide sex linkage.

Silver Laced Sebright Bantam

* See Plate, pages 48 - 49 for examples of main colour or one of colours.

Rosecomb

Type: A true bantam bred from early times, possibly for hundreds of years, and possessing remarkable features in body, wings, ear lobes and comb. A cobby bantam, very small, with very full breast, short back on male, full tail and very decorative comb and face, the whole to be symmetrical.

Origin: British.

Size: Small; 16 to 22oz. (450 to 620g.) the cocks being heaviest.

Comb: Rose with detailed "work" and no large spikes or hollows or ridges. Immaculate with broad main part and tapering to the leader, rising slightly and beyond back of head. Ear lobes must be large, enamel white and round; fine like kid without any blemishes or discoloration.

Tail: Male: Full with curved, broad sickles almost forming a circle; however, should not go too low to complete the circle (45 degree angle). Female: long and upright, neatly folded into a half fan (35^0 angle)

Legs & Feet: Colour black for Black variety; slate in Blues and white in Whites.

Beak & Eyes: Short with slight curve and black, horn or white for Blacks, Blues and Whites respectively. **Eyes:** Brown (Blacks) and reddish bay.

Main Colours:

Black, Blue and White. The Black is the most popular; must have brilliant green sheen.

SINGLE MATING

For breeding a specific characteristic many breeders prefer to breed from a pair of birds rather than a trio or even more hens.This allows much more control to be exercised.

The Felch Chart

In case fanciers believe single-pair breeding is not practical it is necessary to mention the Felch Breeding Chart, which shows that three strains can be developed from one pair of birds. This is reproduced below.

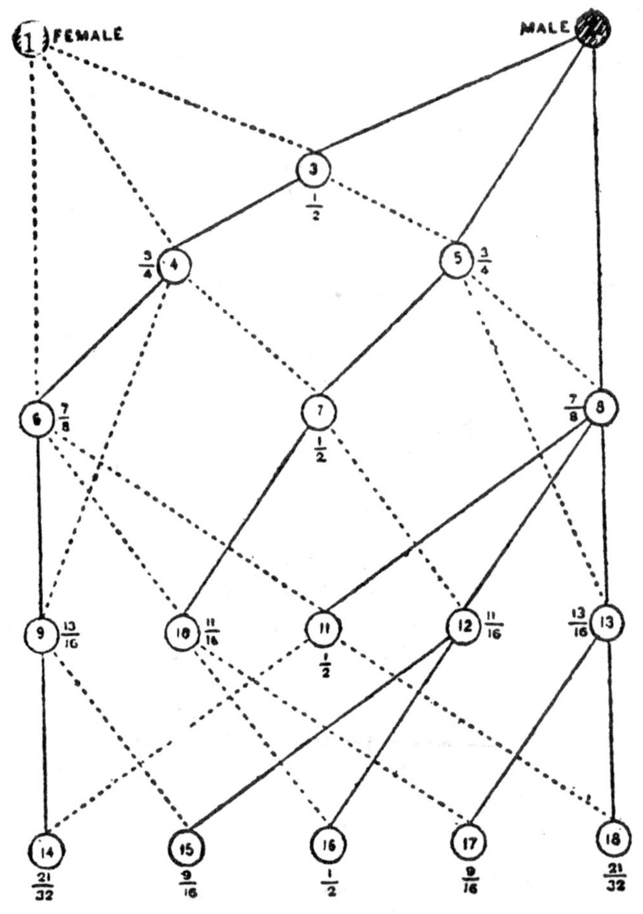

NOTES ON THE CHART

1. The dotted line represents the female.
2. The solid line is the male.
3. Each circle represents the chicks produced. Thus if number 3 is taken this will show that the progeny are 50 % of each of the parents.

There then follows various crosses back, thus changing the percentages from the originals.

Thus: Examples (See Chart for fractions)

Group 3 Females X No 2 Cock = No 5. = 3/4 of Sire No 1.

Group No 3 Male X No 1 hen = No 4. = 3/4 of Hen No 1.

Group No 5 Male X No 4 Pullet = No 7= 1/2 of original pair. This could be a new strain.

Group No 5 X Group No 8 = Progeny = 7/8 of No 2.

Group No 4 Cock X Group No 1 Hen = No 6 = 7/8 of Hen 1 and 1/8 of Cock No 1.

This carries on and uses the blood of the original sire. A cock can now be produced which contains almost 100 per cent of the original hen.

Thus if Group No 6 Male X Group No 4 Female = Group No 9 = 13/16 the blood of Hen No 1.

Yet Group No 9 male X Group No 11 Pullet = Group 14 = 21/32 of Hen No 1.

Features

1. The method allows the genes to be divided and then to be reamalgamated. However, recent research has shown that genes do not work out as expected and there are far more than was originally thought.

2. Dividing in this way is logical, but obviously does not always work out in the precise way suggested by the fractions.

3. The method is difficult to understand and requires specific examples to be worked out and careful records kept.

OTHER FACTS ON COLOUR

Besides the colour of the plumage the fancier should be interested in the colours of the 'other parts'; ie, features such as combs, legs, skin, ear lobes and eyes.

Combs

Combs are of many types - single, rose, pea, and others, but the main colour for the comb itself and the face is red. Generally, the brighter the colour the better because this is an indication of health.

Other natural colours which exist are:

1. **Mulberry**; the Silkies and Sebrights are the best examples. There is a tendency for this colour to be changed by crossing with red, which should be guarded against, like all other alien crosses.
2. **Gypsy**, which is a dusky colour, yet with an obvious vitality behind the darkness. Old English Game such as Black Breasted Black Reds show the colour to the maximum; Brown Reds also show it, although some strains have a red comb. The other main breed is Sumatra Game which has black, gleaming plumage, indicating that the presence of a Melanistic gene is the clue to its existence.

A purple comb on a bird which should have a normal red colour is a serious fault because it often indicates a physical condition such as a weak heart, respiratory trouble or other defect. Sometimes it can be brought on by lack of water.

Eyes

Eye colour is related to a number of factors - down colour and skin. The colour comes from pigment in the iris.

There are many eye colours and these may be listed as

follows:

1. Pale yellow or pearl, often referred to as 'Daw'. The Malays, Aseel and Indian Game have this colour.

2. Brown in various shades seen in dark coloured birds.

3. Red of varying shades seen in OEG and many other breeds.

4. Black which is found in very dark coloured breeds such as Gypsy faced Blacks (OEG). The very brilliant purple pigment in plumage is usually found with a black eye colour and those with beetle-green plumage as seen in the Modern Langshan. This is an elusive colour, but a pale eye on a very dark plumaged bird looks wrong and must be eradicated in the breeding pen.

Ear Lobes

The genetics for ear lobes are apparently quite complex, a number of genes being involved. With exhibition birds which require white ear lobes the continued transfer of the correct type and colour is very important.

Usually all birds in a particular breed which has white ear lobes must have this characteristic. In fact, if taken further it will be seen that the main Mediterranean breeds have ear lobes in varying sizes, all of which are enamel white. This also extends to the varieties within the breed.

Ancona, Andalusian, Leghorn, Minorcas, Spanish, have the white ear lobes, but not the Sicilian Buttercup or Flowerbird, although these are not usually regarded as being true Mediterranean breeds.

Many breeds have red ear lobes, including Brahmas, Plymouth Rocks, Wyandottes, Rhode Island Reds and many others. OEG sometimes show white ear lobes, but this is not consistent and therefore must be regarded as incorrect.

Legs

The leg colour is certainly related to the plumage. Those birds with white skin often have white legs so this is another factor. Strictly, when discussing colour of the legs we mean shanks and not the thighs which are covered in feathers anyway.

There are two genes which determine white leg colour; that is, they inhibit the development of pigment. This also applies to other colours such as blue, slate or grey. Yellow in the shanks is due to the presence of lipochrome pigment (animal fat) in the epidermis* (this also causes yellow skin). With willow legs these are caused by the yellow pigment plus melanic pigment in the dermis.

Black shanks are due to melanic pigment in the epidermis and possibly also the dermis.

The presence of lipochrome is an important indicator of purity of a breed because this causes the bottom of the feet to be yellow; when there is no lipochrome the soles are white. In deciding purity of blood the colour of the soles can be an indication.

***Note:** Because the Victorians did not have the required knowledge on genetics some rather bad mistakes were made when the standards were drafted. Take for instance, the Black Leghorn which is required to have yellow legs; yet these are not the natural colour for black plumage which would normally be black or grey.*

***Epidermis = Outer layer of the skin, whereas Dermis is the inner layer.**

A DICTIONARY OF TECHNICAL EXPRESSIONS USED IN CONNECTION WITH POULTRY

AMATEUR - "One who cultivates a study or art for the love of doing so, and not for gain" - not a professional. In poultry circles it is frequently employed as an equivalent to "novice" - a new hand, a beginner.
BARRED, BARRING Stripes of light and dark across a feather appearing black, white, black (or blue) in an alternate fashion. Barring is usually regarded as being more exact in markings than **Cuckoo.**
BARRED TO THE SKIN - A term frequently applied to the marking on a "Barred Rock" implying the barring is continued from the tip to the root of the feather.
BEAN - The bean-shaped patch of black on the tip of a duck's bill, particularly noticeable in Rouens and Indian Runners.
BEARD - A formation of feathers around the throat of some breeds, as Houdans, Polish, and Faverolles. Related to the term 'Muffles' which is a description for beard and muffs.
BEEFY - A term applied to a large, coarse or overgrown comb.
BEETLE BROWED - overhanging skull formation above eyes (heavy eyebrows) seen in Indian Game or Malays.
BIB - Another expression for beard. (*See* Beard)
BLOCKY - A term designating a thick, square-set bird, with legs well apart, in distinction to a tall, narrow specimen.
BRASSY - The brass-like appearance of the feathers of a fowl's back after being weathered or exposed to the sun. OEG has a variety **Brassy Backed.**
BREED - Any variety of fowl in all its distinct characteristics; usually designated by the outward shape - the TYPE. The breed includes all the **varieties** of colour which are found in it.
BROOD - The family of chickens under one hen or brooder.
BROODY - Desiring to sit or incubate.
CAPON - A male bird deprived of generative organs or a bird being fattened. Caponizing is not practised much now and in some countries is no longer permitted.
CARRIAGE - The way a bird stands, holds itself, and walks.
CARUNCULATED - Having caruncles. A naked fleshy excrescence on the head or neck of birds. Particularly noticeable in turkeys and Muscovy ducks.

CHICK – A newly-hatched fowl. Used only until the bird is a few weeks old.

CHICKEN – A fowl under a year old. For exhibition purposes a bird hatched in the year the show is held, or late in the previous year.

CLODDY – Heavy, thick set.

CLOUDY – Indistinct.

COCKEREL – A young cock. For exhibition purposes a male bird hatched in the year the show is held or late in the previous year.

COCKEREL-BRED – A term to denote that the bird in question was bred from a pen of fowls mated to produce exhibition cockerels. (See Double Mating)

COMB – The fleshy portuberance on the top of a fowl's head.

CONDITION – The state of the fowl as regards health and beauty of plumage .

COVERTS – *See* TAIL COVERTS and WING COVERTS.

CRAW – Same as CROP.

CREST – Formation of feathers on the head in the form of a cap seen on Polands and other crested breeds. The same as TOP-KNOT although this is also used to denote *Tassel.*

CROP – The bag or receptacle for a bird's food immediately after it is eaten, before digestion.

CROP-BOUND – Congestion of the crop so that food cannot pass to the gizzard.

CUCKOO – Barring on feathers, but somewhat indistinct so that the bars are not very distinctly marked as in Barred Plymouth Rocks. Term applies to such cuckoo breeds as Marans, North Holland Blues and Leghorns.

CUP COMB – Shaped like a cup such as found in Sicilian Buttercup or Flowerbird.

CUSHION – The mass of feathers under the tail-end of a hen's back, covering the tail, chiefly developed in Cochins.

DAW EYED – Light coloured eyes, usually Pearl or light yellow seen in Aseel and other Asian breeds.

DEAF-EARS – Another name for ear-lobes

DIAMOND – A term used by game fanciers to denote the wing-bay.

DUBBING – Cutting off the comb, wattles, etc., so as to leave the head smooth and clean. Used to avoid frostbite, but is also standard practice for Modern and OEG. when dubbing scissors are used to remove comb and wattles. In the case of modern Game the dubbing is very close.

DUCK-FOOTED – A term of reproach when applied to a fowl, particularly to a Game bird when its hind toe turns in like a duck's.

EAR-LOBES – The kid-like formation of skin usually white or red in a specific shape and positioned just below the ears. They vary in colour in different breeds, between red, white, blue and cream, and also greatly in size.

FACE - The bare skin around the eye. COARSE FACE is when the skin is wrinkled or puckered, instead of being smooth. GIPSY-FACE is one covered with short black hairs or feathers, giving it a dark appearance or a smooth face which is quite dark, like mulberry, but quite dusky rather than purply.

FLAT SHINNED OR SHANKED - Shape of front of leg found on Aseel, but incorrect on many other breeds when rounded shanks are expected.

FLIGHTS - The primary feathers of the wing, used in flying, ***but tucked under the wing out of sight when at rest.***

FLUFF - Mass of fluffy feathers on thighs of Asiatic breeds such as Brahmas. Also the "under feathers" or hidden parts of the soft feathers on the lowerpart of the shanks.

FOLDED-COMB - A hen's comb which partly falls over and then folds back, instead of falling to one side only.

FOOT-FEATHER - The stiff feathers on the feet of Brahmas, Cochins, etc.

FOXY - A brown or rust colour, often a fault as for instance when on an OEG Partridge hen (across the wings).

FRIZZLED - A term applied to a race of fowls in which each feather is curled backwards.

FURNISHED - Assumed the full characters. When a cockerel has obtained his full tail, comb, hackles, etc., as if adult, he is said to be "furnished".

GILLS - This term is often applied to the wattles, and sometimes more indefinitely to the whole region of the throat.

GULLET - The loose part of the lower mandible; the dewlap of a goose.

HACKLES - The narrow feathers on the neck of birds and in the saddle of the cock. The latter are always termed "saddle" hackles, hackles alone having reference to the neck feathers.

HEN-FEATHERED OR HENNY - Resembling a hen in the absence of sickles of hackle-feathers, and in plumage generally. Applies to OEG Hennies and various other breeds such as Sebrights and Campines, although the latter are not always properly hen-feathered in the sense of sickl;es without curves.

HOCK - The joint between the thigh and shank. Certain breeds, such as OEG must have a bend at the hock, giving the bird a slightly stooped appearance , but with a constant state of alertness.

HORN COMB - Resembling horns or spikes such as found on Sultans or La Fleche.

IDEAL - Technically a drawing or painting representing the perfect bird of any variety.
INBREEDING - Breeding from stock related in blood to each other; eg, Mother to son.
KEEL - The vertical part of the breast-bone. Also applied to independent flesh and skin below the latter.
KNOCK-KNEED - The hocks standing near together.
LACED, LACING - A stripe or edging all round a feather, of some colour different from its ground colour. May be single laced as in a Sebright or Double laced as in an Indian Game hen or on a male *and* female Doubled Laced Barnevelder.
LEADER - The spike terminating the back of a rose-comb., which may be long or short depending onthe breed.
LEAF-COMB - A comb, shaped like a leaf, as in the Houdan.
LEG - In a live fowl this is the scaly part or shank or the In a bird dressed for table, on the contrary, the term refers, as is well-known, to the joint above and known as the drumstick.
LEG-FEATHERS - The feathers projecting from the outer side of the shanks in some breeds, as Cochins and Silkies.
LINE-BREEDING - Breeding conducted on scientific lines.
LOBES - See EAR-LOBES.
LOPPED COMB - Not straight, falling over one side when it should be upright - a fault.
MANDIBLES - Two parts of beak or bill; upper and lower mandibles.
MARKINGS - A general term denoting lacing, barring, pencilling, etc. or other pattern on the feathers.
MEALINESS - A defect in the colouring of feathers although OEG would not normally be penalized. Having the appearance of "meal" sprinkled over it.
MOONS - SPANGLES - A term used when referring to the "half-moon-shaped spangles" on the tail feathers of Spangled Hamburghs.
MOONIES - An ancient breed related to Hamburghs, and possibly one of the first pure breeds to be shown, but now apparently extinct.
MOSSY - Indistinct colour in the markings or smudges.
MOTTLING, MOTTLED - A small tip of white at the end of a feather, e.g. the Ancona when they are known as V-shaped tips. See **SPANGLES.**
MUFF, MUFFLING - Side whiskers of feathers, particularly noticeable in Houdans and Faverolles. The *muff* should not be confused with the beard.
NOVICE - A beginner. As a fancier - one who has not done much winning. Many specialist clubs have their own exact definition of the word, these definitions vary, and when competing in such classes it is advisable to ascertain the exact meaning of the authorities.
PEA-COMB - A comb with three lines of very small spikes, the middle one being highest.
PEN - A cage at a show to hold an exhibit; often used for show-training of birds. An enclosed space for a number of fowls to breed in. The fowls themselves mated together for breeding purposes.

IDEAL - Technically a drawing or painting representing the perfect bird of any variety; the drawings by Ludlow represented the Victorian Ideals and those by Charles Francis a more recent interpretation.

INBREEDING - Breeding from stock related in blood to each other.

KEEL - The vertical part of the breast-bone. Also applied to independent flesh and skin below the latter.

KNOCK-KNEED - The hocks standing near together.

LACED, LACING - A stripe or edging all round a feather, of some colour different from its ground colour.

LEADER - The spike terminating the back of a rose-comb.

LEAF-COMB - A comb, shaped like a leaf, as in the Houdan.

LEG - In a live fowl this is the scaly part or shank. In a bird dressed for table, on the contrary, the term refers, as is well-known, to the joint above and known as the drumstick.

LEG-FEATHERS - The feathers projecting from the outer side of the shanks in some breeds, as Cochins.

LINE-BREEDING - Breeding conducted on scientific lines.

LOBES - See EAR-LOBES.

MARKINGS - A general term denoting the lacing, barring, pencilling, etc., of the plumage.

MEALINESS - A defect in the colouring of feathers. Having the appearance of "meal" sprinkled over it.

MOONS - SPANGLES - A term used when referring to the "half-moon-shaped spangles" on the tail feathers of Spangled Hamburghs.

MOSSY - Confused or indistinct colour in the markings.

MOTTLING, MOTTLED - A small tip of white at the end of a feather, e.g. the Ancona.

MUFF, MUFFLING - The tuft of feathers on each *side* of the face, particularly noticeable in Houdans and Faverolles. Also called *whiskers*. The *muff* should not be confused with the beard.

NOVICE - A beginner. As a fancier - one who has not done much winning. Many specialist clubs have their own exact definition of the word, these definitions vary, and when competing in such classes it is advisable to ascertain the exact meaning of the authorities.

PEA-COMB - A triple comb, resembling three small combs lying side by side, the centre one being the highest.

PEN - A cage at a show to hold an exhibit. An enclosed space for a number of fowls to breed in. The fowls themselves mated together for breeding purposes.

PENCILLING, PENCILLED - Small markings or stripes over a feather. They may run either straight across, as in Hamburghs, or in a crescentic form, as in Partridge Cochins. Also refers to the barred marks on Gold Pencilled Hamburghs as well as the fine markings on OEG hens and certain other breeds with Partridge colouring.

PILE – The name given to a variety of Game and Leghorn fowls which is a white background colour and red markings. Probably a corruption of the word "pied". There is a gene which inhibits the black in Black Reds and changes the colour to white.
POULT – A young turkey.
PRIMARIES – Another name for flight feathers. See "Flights".
PULLET – A young hen under a year old. For exhibition purposes a female hatched in the year the show is held.
PULLET-BRED – A term to denote that the bird in question was bred from a pen of fowls mated to produce exhibition pullets.
REACHY – Tall, upstanding, able to reach high, as in Modern Game and Modern Langshans.
ROACH-BACK – A hunchback, a bird with a faulty spine.
ROOSTER – An American term for cock or cockerel.
ROSE-COMB – A broad, solid comb, the top of which is almost flat and covered with small points. It becomes broader as it recedes from the point and then narrows, terminating in a spike or "leader", which may be short or long.
SADDLE – The rear part of the back, extending to the tail, in a cock, and known as the Cushion in a hen.
SADDLE-HACKLES – *See* "Hackles"
SAPPINESS – The difference in colour in a growing feather to that attained at maturity, due to the presence of "sap". The expression is more generally used (though incorrectly), to denote a yellow tinge in the plumage of white birds after the feathers have matured.
SECONDARIES – The main flight feathers of the wing, which are visible when the bird is at rest.
SELF-COLOUR – A uniform tint over the feather of the bird.
SHAFT – The middle stem of a feather, known as the quill or shaft.
SHAFTY, SHAFTINESS – A term denoting the shaft of a feather is too noticeable, due to its colour differing from that of the web of the feather.
SHANK – The scaly part of the leg.
SHEEN – That which shines. "Sheen" denotes the condition of a bird, being dull when out of health and bright or shiny when in condition. In some breeds the sheen is beetle green, whereas in others it is steel-blue.
SHOULDERS – Part near the neck which would include the shoulder part of the wings known as the Wing Bows.
SICKLES – The top curved feathers of a cock's tail. The lesser sickles are those between the true sickles and the tail-coverts.
SIDE-SPRIGS – Fleshy growths at the side of a single-combed bird.
SILKIE – A small fowl, large and bantam, which has furry feathers and dark coloured flesh.
SINGLE-COMB – A single layered comb, having its spikes in line behind each other. These spikes or serrations should be wedge-shaped; their average number is six, one less being preferable to a greater number. With some breeds; eg, the Leghorn, the comb in the females falls gracefully to one side.

SLIPPED WING - Primary feathers do not hold up properly when bird standing, so there is an obvious gap.
SMOKY UNDERCOLOUR - Dark colour of fluff, usually regarded as a defect in some breeds.
SMUTTY - Indistinct in colour. Particularly applied to "barred" varieties.
SPANGLING - The marking produced by each feather having a large spot of some colour, differing from the ground-colour. Usually applies to OEG Spangles, but also covers Speckled Sussex. In Anconas the marks are known as V-shaped tips.
SPLIT WING - A larger gap than normal between the secondaries and primary feathers; any wide gap in feathers except when moulting.
SPORT - An unexpected offspring. A reversion to a distant ancestor. A new variety.
SPUR - The sharp, offensive weapon near the heel of a cock or hen. If it grows very long it should be removed.
SQUIRREL-TAILED - A bird is so called when its tail projects over the back and closely approaches the head at an angle of above 90 degrees.
STAG - An alternative term for cockerel , chiefly used by game breeders.
STANDARD - An abbreviation for *Standard of Perfection* -- the description issued by the Poultry Club or by Specialist Clubs of any particular variety or breed.
STRAIN - A race of fowls, which, having been carefully bred by one breeder, or his successors for years, has acquired and established an individual character of its own.
STRIPED - The condition where a streak or line of a different colour extends length wise through the feathers, usually seen in the hackle and occasionally in the saddle.
SURFACE-COLOUR - The colour pertaining to the exposed feathers of a fowl. *See* "Under-Colour"
SYMMETRY - Perfection in the proportions of a fowl; balance in OEG.
TAIL-COVERTS - The soft, glossy, curved feathers at the sides of the bottom of the tail and below the sickle feathers.
TAIL-FEATHERS - These are the straight and stiff feathers of the tail only, which grow inside the sickles and tail coverts of the cock.
THIGH - The first joint above the shank. In cooked poultry it is referred to as the "leg" or "drumstick"
THUMB-MARKED - A term applied to the single comb of a cock having an indentation on one side and a corresponding projection in the other as though caused by the pressure of the thumb.

TICK - A speck of colour found in the fluff of feathers of whole and buff-plumaged birds.
TOP-KNOT - An alternative name for "Crest" or tassel.
TRI-COLOURED - A term applied particularly to Buff varieties, when instead of one shade they possess darker hackles and saddles and yet darker wing-bows.
TRIO - A cock or cockerel with two hens or pullets.
TYPICAL - Expressions denoting nearness of perfection in the shape of the bird.
TYPE - Standard shape and appearance of pure breed.
VARIETY - A definite division of a breed. Such divisions are generally caused by distinctive colourings, but may result from other peculiarities, *e.g.* Single-comb Rhode Island Reds and Rose-comb Rhode Island Reds.
VULTURE-HOCK - Quill feathers which stand out above the hock-joint. Except on Sultans and Booted bantams, a defect, but the feathers must be both stiff and projecting to be condemned.
WALNUT COMB - Resembles a walnut cut across its full length and then laid on its edge. Also resembles a Strawberry comb.
WASTER - A bird unfit for breeding or exhibition or production. Only fit for killing purposes.
WATTLES - The fleshy structures at each side, below the base of the beak, more pronounced and developed in the male sex. Non-existent in some breeds; eg. Aseel and very small in others. Cut off when dubbing.
WEB - Applied to the feather, it is the flat or plume portion; the web of the foot is the flat skin between the toes, whilst the web of the wing is the triangular portion of skin observable when the wing is extended.
WHEATEN - The colour of the female in Black-breasted Red Game and other breeds - originally a rich uneven creamy-brown resembling the appearance of a field of ripe wheat , but now a *light* cream, as even as possible.
WHISKERS - *See* "Muff, Muffling".
WILLOW - A greenish-yellow of various shades, appreciated as leg-colour in Black-red Game but a defect in breeds which should possess yellow legs.
WING-BAR - The centre portion of the wing, covered by the wing-coverts. In most varieties the "bar" is a distinct colour from the other parts of the wing. It should be perfectly even and clear.
WING-BAY - The end of the wing, triangular in shape, formed by the lower end of the secondary flight feathers. "Diamond" is an alternative name.

WING-BOW - The upper section of the wing.
WING-BUTTS - The corners or ends of the wing.
WING-COVERTS - The broad feathers covering the roots of the secondary quill feathers.
WORK - A term applied to the small points on the almost flat top of Rose-combs.
WRY-TAIL - A tail carried on one side and therefore not straight in line with the body.

INDEX Also see List of Breeds on page viii (Prelims)

NOTE Colour Plate List in Prelims.
Colour Plates between pages 48/49.

Indian Game 25
Javas 7
Jungle Fowl 12, 13